EINSTEIN
1905

JOHN S. RIGDEN

EINSTEIN
1905

The Standard of Greatness

Harvard University Press
Cambridge, Massachusetts
London, England

The illustration on p. 70 is from Mary Jo Nye, *Molecular Reality* (Elsevier, 1972), p. 11 and is reprinted with the permission of Mary Jo Nye. Photo on p. xii is courtesy of the Archives, California Institute of Technology. Photos on pp. 18 and 106 by Robert Sears / courtesy Caltech Archives. Photos on pp. 42, 56, 74, and 126 are courtesy of AIP Emilio Segrè Visual Archives (p. 42, National Archives and Records Administration; p. 126, Gezari Collection). All photos are reprinted with permission.

Library of Congress Cataloging-in-Publication Data
Rigden, John S.
Einstein 1905 : the standard of greatness / John S. Rigden.
p. cm.
Includes bibliographical references and index.
ISBN 0-674-01544-4 (cloth : alk. paper)
1. Einstein, Albert, 1879–1955—Influence.
2. Physicists—Intellectual life.
3. Quantum theory.
I. Title.

QC16.E5R54   2005
530.11—dc22   2004054049

Contents

Preface

This book celebrates Albert Einstein's 1905. In six months Einstein wrote five papers that deeply influenced the course of twentieth-century science. These papers from the hand of a then-unknown physicist make 1905 one of the most memorable years in the history of science and, without doubt, make the six months from March 17 to September 27 the most productive six months any scientist ever enjoyed. Einstein's 1905 papers spoke for themselves then and now, and they have influenced a vast body of physical research during the intervening decades.

In any celebration the emphasis is on the positive. Einstein has grown to almost mythic proportions, which challenges any author writing about him to avoid expanding the myth by making more of him than is justified, but also to avoid contracting the myth by yielding to the seductive temptation to bring greatness down to size. Both giant makers and giant killers respond to their respective tasks with enthusiasm.

As a celebration of Einstein's 1905, this book is not about Einstein the man or about his generous or petty deeds except as the personal characteristics of the man are reflected in *what* he did in 1905 and *how* he did it. Einstein had a way of working and a way of thinking that determined not only what he chose to think about, but also how he thought about it. As a physicist there are few who can compare.

Einstein's physical intuition was both accurate and powerful. As a result, he saw significance and meaning that others could not see in common, everyday observations. From quaint images—an object falling freely toward Earth or a speedster running alongside a beam of light—came breathtaking new insights into the reality behind Nature's appearances. As new physical concepts emerged from his work, Einstein exercised a prescience that defies description. For example, Einstein recognized and, in essence, predicted nuclear energy thirty-four years before the discovery that made it possible. He predicted gravitational red shift at least forty-four years before it was confirmed. Einstein envisaged stimulated emission thirty-seven years before light was amplified by means of stimulated emission in the laser, and predicted a new state of matter, the Bose-Einstein condensate, seventy years before it was discovered. Finally, Einstein described entangled quantum states over thirty years before they became an active subject of physical research. The sluggish pace with which some of Einstein's ideas were accepted and the years that passed before his predictions were verified suggest that he was often working decades ahead of the times.

There are a few who promote the thesis that Einstein's papers had an unnamed author; namely, Einstein's wife, Mileva. The views of such individuals range from Mileva as the principal intellectual force to Mileva as an active intellectual co-contributor. The evidence for either thesis is scant and inferential; it is based primarily on love letters written by Einstein to Mileva during the years immediately preceding their marriage in 1903. Across the field of Einstein scholars, this thesis is almost universally rejected. In what follows I assume Einstein to be the author of the 1905 papers that appear under the name A. Einstein. The stylistic similarities between these papers and his later ones strongly suggest a common author.

For advances in science, the twentieth century is peerless. Old windows onto the natural world widened until they approached natural limits. New windows were opened. How can we explain the sudden acceleration of discoveries in the twentieth century? An unusually large number of first-rate minds numbered among the vast population of scientists and drove the exponential growth of science. Many of these scientists already occupy an honored place in history. Yet no one has approached the quantity and quality of work that Einstein produced in March, April, May, June, and September of 1905.

In 1905, the outpouring of one man's genius changed forever our understanding of Nature, breaking with the past and establishing new territory for the science to follow. Einstein wrote twenty-one reviews and five papers in that year, one of which was his dissertation. Einstein had a long career, but he never had another year the equal of 1905. Not many of his other scientific papers compare to the high standard he established for himself in that year. One exception is Einstein's general theory of relativity (1916), which in both content and style is considered a masterpiece. In the Epilogue, I discuss briefly five post-1905 papers, including the general theory.

EINSTEIN
1905

Albert Einstein in the Bern patent office. This picture was taken at the height of Einstein's creative outburst when he wrote five time-less papers in six months. These papers set the table for much of twentieth-century physics.

The Standard of Greatness: Why Einstein?

Einstein had a rare ability to recognize the core principles that account for the world we observe. His perceptions were never diverted by the many fascinating distractions that enshroud Nature's underlying reality. Common sense, often a source of comfort, can distract. Experimental data, always the ultimate authority, can distract. Einstein saw beyond common sense and, while he respected experimental data, he was not its slave. He saw Nature as it is.

Albert Einstein and Isaac Newton are routinely identified as the two greatest physicists of all time. To be sure, this is a distinction of the highest order, but how can such an unquantifiable, abstract distinction be justified? If "all time" were reduced to "our time," the abstractness might dissipate somewhat; therefore, let us exit "all time" and enter the twentieth century. On December 31, 1999, Einstein was named "Person of the Century" by *Time* magazine. We can scan the faces of the extraordinary people who put their stamp on the century just ended: from the arts to the sciences, from world leaders to great writers, we have lived among many whose presence defined the century. It remains a challenge, perhaps even a greater challenge because they are so familiar, to appreciate the significance of Einstein being given precedence.

We might smile at the folly of naming one physicist as best or naming one man or one woman person of the century. We might agree or disagree with the ranking accorded Einstein in "all time" or in "our time." Nevertheless, in our era, Einstein is the standard by which we popularly judge intellect.

It is often difficult to assess famous people fairly. However, Einstein's stature derives from his work, and his work is there for all to see and judge. Were it not for the profundity of Einstein's many accomplishments, were it not for the fact that his work, even after a hundred years, continues to be the prism with which contemporary scientists seek to open our universe to complete understanding, then Einstein would merely be another great scientist. Einstein's greatness rests upon the *what* and the *how*. The consequence of the *what* cannot be exaggerated; the applause given the *how* cannot be overdone. At the age of twenty-six, he wrote five papers that changed science forever. According to Einstein himself, "A storm broke loose in my mind."

What Did Einstein Do?

Until his death, physics consumed Einstein's waking hours. His professional career began December 13, 1900, when he finished his first scholarly paper. His career ended on the day he died, Monday, April 18, 1955, at 1:15 in the morning. The day before death stopped his work, he asked that his writing materials be brought to his hospital room so he could continue his work. Those writing materials were by his bedside awaiting his hand Monday morning. Had he lived another day, his career would have been one day longer.

Einstein's career covered fifty-four years, but his scientific reputation and scientific immortality rest almost entirely upon the work he did during the first twenty-five. The development of quantum mechanics, started in 1900 by Max Planck and completed in final form in 1927, has dominated physics ever since. In 1927, quantum mechanics came to rest on both physical and philosophical principles about the nature of physical reality that Einstein could not accept. For the last thirty years of his life, Einstein did little that actively contributed to the frontiers of physics. But by 1925, when he was moving out of the physics mainstream and into what became his life-long obsession to unify gravitation and electromagnetism, Einstein's standing in the community of physicists was unchallenged. His advice, his thoughts, and his judgments were sought avidly by all leading physicists and by those mainstream physicists who were actively shaping the discipline, physicists like Niels Bohr and Werner Heisenberg, who were deeply troubled by the doubts Einstein brought to their work. Because Einstein's ideas always had to be taken seriously, his influence on the discipline was powerful.

So *what* did Einstein do? In a letter to his friend, Conrad Habicht, written in mid-May 1905, Einstein begins an answer to the question:

> I promise you four papers . . . the first of which I might send you soon, since I will soon get the complimentary reprints. The paper deals with radiation and the energetic properties of light and is very revolutionary, as you will see . . . The second paper is a determination of the true sizes of atoms from the diffusion and the viscosity of dilute solutions of neutral substances. The third proves that, on the assumption of the molecular [kinetic] theory of heat, bodies of the order of magnitude of 1/1,000 mm, suspended in liquids, must already perform an observable random movement that is produced by thermal motion; in fact, physiologists have observed (unexplained) motions of suspended small, inanimate, bodies, which motions they designate as "Brownian molecular motion." The fourth paper is only a rough draft at this point, and is an [sic] electrodynamics of moving bodies which employs a modification of the theory of space and time; the purely kinematic part of this paper will surely interest you.[1]

Einstein did not tell Habicht about his fifth paper. These five papers, completed between mid-March and the end of September, were all published in the leading German physics journal,

Annalen der Physik. Three of the papers—the March paper on the particle nature of light, the May paper on Brownian motion, and the June paper on the special theory of relativity—are universally regarded as epoch-making papers. The April paper, Einstein's doctoral dissertation, receives little attention, although it is one of Einstein's most cited papers and set the stage for the May paper. The September paper, in which the famous equation $E = mc^2$ first appeared, followed as an unanticipated consequence of the June paper. The five papers touch foundational issues in separate areas of physics.

Although Einstein wrote these five papers over a six-month period, it would be misleading to suggest that in the spring of 1905, a flood of profound ideas suddenly popped into his mind. Indeed, Einstein had been ruminating about these ideas for some years, until the spring of 1905, when he began furiously writing papers.

With some frequency, new scientific ideas are created, received with interest, and remain until they are discarded when their validity is compromised or their usefulness wanes. By contrast, what Einstein did in 1905 stands firm to the present day. After a century of rapid advances in all areas of science, Einstein's work remains solid, and the little equation $E = mc^2$ has become an icon of science.

What Einstein did in 1905 has impacted not only twentieth-century physics but also science in general, actively shaping subsequent scientific endeavors. Two revolutions occurred in physics during the twentieth century. Both had broad and deep implications. The first was the result of Einstein's June paper on special relativity, which required the restructuring of our ideas about space and time, the most basic concepts of physics. This revolution was completed in 1915, again by Einstein, with his general theory of relativity. The second revolution was quantum mechanics, which took form between 1925 and 1927. Einstein's March paper is a pillar supporting the edifice of quantum mechanics.

Finally, the April and May papers brought statistical fluctuations into statistical physics and influenced that field's subsequent development.

The general public is largely unaware of what Einstein did in his March, April, and May papers, but the June paper that presented to the world the special theory of relativity touched the general public in uncommon ways. People experience space and time. People live in space and time. Yet, the ideas of space and time are intangible and mysterious. In June 1905, Einstein essentially restructured our ideas about space and time and the logical consequences of his theory violated our common sense in the harshest of ways. Nonetheless, these bizarre consequences have been repeatedly verified by experiment and they are accepted as true.

Einstein's identity is strongly linked to his June and September papers. It is probably fair to say that most members of the general public do not understand how the concepts of space and time were changed by these papers, but people do understand that Einstein did something enormously profound.

Einstein's most notable and his single most important scientific contribution was the general theory of relativity, which he completed in 1915. In the years immediately following 1905, Einstein thought about how Newton's gravitational theory could be brought under the umbrella of the special theory of relativity. In 1907, he had what he called "the happiest thought of my life."

> I was sitting in a chair in the patent office at Bern when all of a sudden a thought occurred to me. "If a person falls freely he will not feel his own weight." I was startled. This simple thought made a deep impression on me. It impelled me toward a theory of gravitation.[2]

As the story goes, Newton saw an apple fall as he sat in his Woolsthorpe orchard and his gravitational force law followed.

Einstein thought of a person falling as he sat in his Bern patent office and his general theory of relativity followed.

Einstein's general theory of relativity is a theory of gravitation. His insight that if a falling person "dropped" an apple, the falling person would see the apple not falling but remaining stationary at his or her side led eventually to the conclusion that the gravitational force is a consequence of the curvature of space. But this insight did not come easily; it was a long struggle for Einstein. He described his effort to a colleague, Arnold Sommerfeld, in Munich: "I am now working exclusively on the gravitational problem . . . [O]ne thing is certain: never before in my life have I troubled myself over anything so much, and I have gained enormous respect for mathematics, whose more subtle parts I considered until now, in my ignorance, as pure luxury! Compared with this problem, the original theory of relativity [the special theory] is child's play."[3] Einstein's concentrated efforts on the general theory ended in 1915 when he completed what is regarded as his single greatest achievement.

On November 6, 1919, at the joint meeting of the Royal Society and the Royal Astronomical Society held in London, the results were announced of a May 1919 expedition whose purpose was to test a particular prediction of Einstein's general theory. Arthur Eddington had traveled to the island of Princípe, an island off the west coast of Africa, to observe a total eclipse of the Sun. At the moment of totality, Eddington observed that the light from a distant star deviated from its straight-line course of motion as it passed by the darkened Sun. One prediction of general relativity, that a massive object should influence the motion of light, was verified. The next day, the London *Times* carried the headline, "Revolution in Science. Newtonian Ideas Overthrown." Almost immediately, Einstein was a world celebrity.

The confirmation of Einstein's general theory came at the end of the Great War. The public, tired of four brutal years of conflict, embraced Einstein as the epitome of humanity's noble side. By the power of pure thought, Einstein had divined deep realities of Nature and had enhanced our understanding of the universe in which we live.

In the seventeenth century, Galileo and Newton built the foundations of an intellectual revolution. After Galileo and Newton, we saw our world in a new way and the methods of science were changed forever. In the century just ended, the way scientists conceptualize the world changed again; this time, with two revolutions. Einstein, with his special and general theories of relativity, was the sole architect of one revolution. Quantum mechanics, the second revolution in scientific thought, had several architects, Einstein among them. Through these theories, relativity and quantum mechanics, Einstein will influence the pursuit of scientific knowledge into the distant future.

Einstein accomplished much in the first twenty-five years of his professional career, and the five seminal papers he wrote in six short months in 1905 represent a major part of that. But Einstein's greatness is also a direct consequence of how he did what he did. How he worked and how he approached his work enhance even more his singular accomplishments.

How Did Einstein Do It?

A careful examination of Einstein's 1905 papers demonstrates both his discrimination in selecting the specific problems he would study as well as his unusual approaches to those problems. Einstein once said, "I want to know how God created this world . . . I want to know His thoughts, the rest are details."[4] His comment reveals the kind of problems that attracted Einstein, those great problems that were akin to God's thoughts. He eschewed trivialities; rather, he sought to know what was in the Maker's mind when He put it all together. Einstein expressed the same general idea in another way: "What I'm really interested in is whether God could have made the world in a different way; that is, whether the necessity of logical simplicity leaves any freedom at all."[5] And because Einstein had the ability to see the Nature we experience through the unseen basic principles that govern our experiences, he was, as the great American physicist I. I. Rabi used to say, "walking the path of God."

In his 1902 book *La Science et l'Hypothèse*, the mathematical physicist Henri Poincaré identified three fundamental yet unsolved problems. One problem concerned the mysterious way ultraviolet light ejects electrons from the surface of a metal; the second problem was the zig-zagging perpetual motion of pollen particles suspended in a liquid; the third problem was the failure of experiments to detect Earth's motion through the ether. In 1904, Einstein read Poincaré's book.[6] He had also been thinking about these problems, independently of Poincaré. For Einstein, they were clearly part of God's thoughts. One year later, in 1905, he solved all three.

Einstein's approach to physics was both powerful and distinctive. He was intrigued rather than dismayed by apparent contradictions, whether they consisted of experimental results that conflicted with theoretical predictions or theories with formal inconsistencies. For example, discontinuity and continuity are contradictory. The particle nature of matter is a prime example of discontinuity; the wave nature of light is a prime example of continuity. That matter, by consisting of atoms, was discontinuous was largely accepted in 1905; that light, by consisting of waves, was continuous was a foregone conclusion long before 1905. Particle versus wave; discontinuity versus continuity—the contradiction captivated Einstein. He was driven to unify disparate or contradictory physical ideas and, in the process, to simplify the theories used to represent them.

Einstein was well aware of his ability to see beyond appearances and to visualize the core principles of Nature. In his auto-biography, he compared his intuitive abilities in mathematics and in physics. In mathematics he described himself as "Buridan's ass . . . unable to decide upon any specific bundle of hay . . . my intu-ition was not strong enough in the field of mathematics . . . to differentiate clearly the fundamentally important, that which is really basic, from the rest of the more or less dispensable erudi-tion." In physics, however, "I soon learned to scent out that which was able to lead to fundamentals and to turn aside from everything else, from the multitude of things which clutter up the mind and divert it from the essential."[7]

This self-knowledge gave him self-confidence; in fact, Einstein was supremely confident about his physics. It did not matter that his ideas were sometimes rejected definitively by his most distin-guished colleagues. It did not matter that his line of thinking sometimes ran counter to the lines of thinking that were fashion-able among his influential contemporaries. It did not matter that experiments specifically designed to test his theoretical predic-tions produced data in conflict with his predictions. His response was to wait quietly and patiently. He seemed to know that his colleagues would soon come around or that something was amiss with the design or execution of the experiment. Einstein knew his theory was correct.

Einstein was confident in the correctness of his physics because he developed it in a particular way. His physics was always com-patible with certain general principles which, if false, would im-pact physics in fundamental ways. If the general principles were true, his physical theory had to be true as well. Einstein revealed this brimming confidence when a prediction of his general theory of relativity was tested. After a long wait for the experimental re-sult, from May to November 1919, he finally heard that the ex-perimental measurement had validated his theory, but the an-nouncement elicted little interest from Einstein. When asked why he was not excited by the great news that confirmed his prediction, he responded, "I knew the theory was correct." But, he was asked, suppose your prediction had been refuted? "In that case," answered Einstein, "I'd have to feel sorry for God, because the theory is correct."[8]

Einstein's confidence was not arrogance. Einstein had carefully examined the basic principles in all possible ways, and when the various pieces meshed together flawlessly, he knew the theory had to be correct. A reliance on core principles gave Einstein confidence. Einstein's confidence in his work was further strengthened because he mostly worked alone.

In 1905, Einstein was not at a major university among distinguished physicists. He was not even at a minor university among undistinguished physicists. Einstein was neither at a university nor was he among physicists. He was a junior clerk in the patent office in Bern, Switzerland. In this environment, he was cut off from the larger world of physics; he had no professional physicists with whom he could share thoughts, collaborate, or argue. No one in his office would alert him to an important new paper in physics. Even his access to libraries was limited. In September 1907, two years after Einstein's relativity paper, Johannes Stark wrote to Einstein and asked him to write a paper reviewing the history of relativity theory since 1905. Einstein responded, "I must . . . note that I am not in a position to acquaint myself with *everything* published on this topic, because the library is closed during my free time."[9] Compared to other physicists, Einstein was isolated.

That does not mean he was entirely out of touch. He had read Poincaré's book, and he did have libraries available to him. Yet in his papers he does not cite the current literature as often as he might have. There are no citations at all, for example, in his June paper on the theory of relativity. What Einstein read and what Einstein knew cannot be known definitively and, as a result, scholarly judgments differ on why so few references appear in his papers.

The job at the patent office was good for Einstein. He enjoyed examining applications for patents on technical devices. He was good at working through the details of a patent application. He was fast. He had time to think. On November 17, 1909, after he had resigned from the patent office (in July) and became an associate professor of physics in Zurich with courses to teach, he wrote to his friend, Michele Besso, "I am *very* occupied with the courses, so that my *real* free time is less than in Bern."[10] The quiet of the patent office may have been essential to him in 1905.

Isolated from active working physicists as he certainly was, Einstein became accustomed to working alone. For the most part, he continued to do so throughout his professional career. He was the sole author of the 1905 papers and most of his other papers. Because the approach he took to physics was singular, neither credit nor blame could be shared. With no co-author dabbling in his work and with confidence in his own insights, he had no doubts about the correctness of his work.

Einstein's confidence revealed itself in still another way. C. P. Snow, in conversations with the great mathematician G. H. Hardy, once discussed Einstein. What words could be used to adequately describe this man? Quickly, the words "great," "gentle," and "wise" were agreed upon. But that, at least for Snow, was not enough. In the end, Snow proposed the descriptor "unbudgeable." "Unbudgeable" captures a quality of Einstein's that the words "great," "gentle," or "wise" do not touch upon. Einstein was indeed unbudgeable. For thirty years, Einstein took issue with the tenets of quantum mechanics as they were developed by Niels Bohr, Paul Dirac, Werner Heisenberg, Wolfgang Pauli, Erwin Schrödinger, and others. The best of these attempted to convince Einstein that the probabilistic interpretation of quantum mechanics represented physical reality as it really was. For thirty years, Einstein watched as the successes of quantum mechanics mounted and multiplied. But Einstein was unbudgeable. To the end, he believed that quantum mechanics represented a holding pattern for physics and that some day it would be replaced by a theory that conformed to his causal views of Nature. Confident in his beliefs, which were reflected in his physics, Einstein held fast and waited.

Einstein's stand in regard to quantum mechanics prompted some physicists to conclude that physics had passed him by. At the same time, however, many of Einstein's contemporaries shared an intellectual uneasiness about quantum mechanics. Some important physicists today believe that, in the end, Einstein's views of quantum mechanics may be vindicated. Only time will decide.

Einstein worked at the level of basic principles and he did it alone. That's how he did his work. As a result, he had unshakeable confidence in his results. It is the "what" and the "how" that answers the question "why" . . . why Einstein is the standard of greatness.

Beyond the What and the How

Einstein's personal qualities only added to the mystique that grew to surround him in the years following 1905. There was an other-worldliness about Einstein that made him appear to be above and beyond the concerns of ordinary people. The most obvious manifestation of Einstein's detachment was the bohemian lifestyle he eventually adopted. He walked around with no socks, his trousers were too short, his clothes were rather dumpy, and his long hair rarely saw a comb. His looks were consistent with his home. The house at 122 Mercer Street in Princeton was a simple, modest dwelling. He had little interest in material possessions.

Once Einstein became famous, his offbeat behaviors and his simple lifestyle only enhanced his monumental reputation. Of course, Einstein was well aware of this. He could have altered his mode of dress, but he did not. That's the way he was. Shining through Einstein's strange appearance, even overriding it, was an uncommon intimacy, an attractive warmth that emanated from him. His wispy hair, sticking out in all directions, framed a fascinating face with large, gentle eyes, eyes that commanded attention. Even in his early years, Einstein's large head, according to the astronomer Charles Nordmann, "instantly attracts attention . . . A little mustache, dark and very short, adorns a sensuous mouth, very red, rather big, with its corners betraying a permanent slight smile. But the strongest impression is that of stunning youthfulness."[11] Einstein's quaint appearance together with his celebrity status endeared him to the public.

Einstein's other-worldliness went beyond his physical appearance. Einstein lived the life of a loner. Rudolf Ladenburg, who worked with Einstein in Berlin (and later at Princeton) said, "There were two kinds of physicists in Berlin: on the one hand was Einstein, and on the other all the rest."[12] Partly, Ladenburg was saying that Einstein stood head and shoulders above the other physicists, but he was also referring to Einstein's aloofness. At the University of Berlin, Einstein was a member of the faculty, but he spent little time participating in university life.

In 1930, Einstein described himself as follows:

My passionate interest in social justice and social responsibility has always stood in curious contrast to a marked lack of desire for direct association with men and women. I am a horse for single harness, not cut out for tandem or teamwork. I have never belonged wholeheartedly to any country or state, to my circle of friends, or even to my own family. These ties have always been accompanied by a vague aloofness, and the wish to withdraw into myself increases with the years. Such isolation is sometimes bitter, but I do not regret being cut off from the understanding and sympathy of other men. I lose something by it to be sure, but I am compensated for it in being rendered independent of the customs, opinions, and prejudices of others, and am not tempted to rest my peace of mind upon such shifting foundations.[13]

Independent of the emotional connections that often complicate and even disrupt others' lives, Einstein exuded a sense of contentment that complemented his unkempt appearance.

Einstein's detachment was certainly liberating; he could pursue any problem wholeheartedly for as long as he wished. His skill at concentrating on a problem for years was demonstrated by his work on the general theory of relativity. Einstein's ability to focus on his intellectual life further added to his mystique.

In contrast with this free-spirit detachment, however, was a steadiness that only enhanced the general public's admiration for the man. Einstein was a man of principle and could act with courage. During the Great War, Einstein was one of four scientists who signed a "manifesto to Europeans" in which scientists and artists were criticized for having "relinquished any further desire for the continuance of international relations" and calling "for all those who truly cherish the culture of Europe to join forces."[14] As Abraham Pais, an Einstein biographer, notes, "The period of 1914–1918 marks the public emergence of Einstein the radical pacifist, the man of strong moral convictions who would never shy away from expressing his opinions publicly, whether they were popular or not."[15] Einstein was a man of conviction and, in memorable words, he conveyed his beliefs to the public.[16]

Jacob Bronowski captured both Einstein's modesty and his principled toughness when he wrote: "He was quite unconcerned about worldly success, or respectability, or conformity; most of the time he had no idea what was expected of a man of his eminence. He hated war, and cruelty, and hypocrisy, and above all he hated dogma—except that hate is not the right word for the sense of sad revulsion that he felt; he thought hate itself a kind of dogma."[17]

Einstein attracted attention long before he was famous. It is often pointed out that Einstein was slow in learning to talk and that he was a poor student. Immediately, hope is born in the heart of every fond parent and truant child; even the brightest feel a sudden affinity for Einstein the student. But is it true? Einstein himself acknowledged that his parents were concerned about his long, slow start in talking. The charge of his being a poor student is less certain. If "poor student" means "dumb student," the answer is decidedly no. If "poor student" means "uncooperative student," the answer is probably yes. Einstein was bored by much of the work teachers gave him. As teachers grew impatient with young Einstein, Einstein grew impatient with them. He was often unwilling to complete the dull tasks that his teachers assigned. He was particularly offended by authoritative teachers who required their students to learn through exercises and drills.

Einstein was often rebellious in an authoritative environment. And it cost him. He did not attend his classes, choosing instead to study those things that interested him. When he applied to the Federal Institute of Technology in Zurich, he failed the entrance exams. A year later he applied again, and this time he passed the exams. When he and three other classmates graduated in 1900, they were qualified to teach. His classmates all got teaching positions; Einstein did not. Einstein took a series of temporary jobs until June 16, 1901, when he was appointed technical expert third class at the patent office in Bern, Switzerland.

The rebellious Einstein became a gentle, accommodating Einstein as he matured. The number of people who wanted to see him in his later years was enormous. His secretary and assistant,

Helen Dukas, somehow reduced the number to a tolerable level. Einstein nevertheless saw many of these people and did so gracefully. He received bags of letters and personally answered many of them, especially the letters he received from children. Einstein's personal characteristics, taken altogether, magnified the products of his mind and, in the process, enhanced his stature as a man.

Although Einstein died in 1955, he remains the standard of greatness. Smart kids are often nicknamed "Einstein." "Hey Einstein," we ask the class genius, "what did you get on the test?" When television commentators want to refer to real intelligence, they mention Einstein. Why Einstein? He was certainly smart, but many people are smart. Einstein, however, is more than simply a symbol of intelligence. When Einstein recognized truths about the natural world by pure acts of mind, he exemplified what is best about being human. And when, through it all, he exuded a noble modesty, he entered the consciousness of all people.

On a Heuristic Point of View about the Creation and Conversion of Light

A. Einstein

The paper was received by the editor of the German journal *Annalen der Physik* on March 18, 1905, and was published shortly thereafter in volume 17, pages 132–148.

Albert Einstein and Robert Millikan on February 26, 1931, while Einstein was a visitor at the California Institute of Technology. It was Millikan who beautifully confirmed a consequence of Einstein's March paper yet adamantly rejected Einstein's basic March thesis that produced the consequence he confirmed. The March paper is the only one of his 1905 papers that Einstein himself called "revolutionary."

The Revolutionary Quantum Paper

In the March paper of 1905, Einstein directly challenged the orthodoxy of physics: orthodoxy that had grown and strengthened for over a century; orthodoxy that rested on bedrock experiment and far-ranging theory.

All physicists in 1905 knew what light was. Whether from the Sun or an incandescent light bulb, light was known to be a wave; that is, a succession of equally spaced crests separated by equally spaced troughs where the distance between the crests (or the troughs) determined the light's color. All scientists knew, without doubt, that light originated at a source, spread out evenly and continuously through all the space accessible to it, and propagated from place to place as electromagnetic crests and troughs. Light was called an electromagnetic wave or, more generally, electromagnetic radiation. In 1905, the wave nature of light was an established, incontrovertible fact.

In the face of this universally held knowledge, Einstein proposed that light was not a continuous wave but consisted of localized particles. As Einstein wrote in the introduction to his March paper,

> According to the assumption to be contemplated here, when a light ray is spreading from a point, the energy is not distributed continuously over ever-increasing spaces, but consists of a finite number of energy quanta that are localized in points in space, move without dividing, and can be absorbed or generated only as a whole.[1]

This sentence has been called "the most 'revolutionary' sentence written by a physicist of the twentieth century."[2]

Einstein anticipated the impact of his paper. In May 1905, before the paper appeared in print, he informed his friend Conrad Habicht that a forthcoming paper on the properties of light was "very revolutionary."[3] From a modern perspective, at least three of Einstein's 1905 papers were similarly innovative, but for Einstein in 1905, it was only the "assumption considered here [the March paper]" that represented a sharp break with established tradition. It was revolutionary at the time and it remained revolutionary. In June 1906, the future Nobel Prize–winning physicist Max Laue wrote to Einstein unequivocally denying Einstein's assumption:

> When, at the beginning of your last paper, you formulate your heuristic standpoint to the effect that radiant energy can be absorbed and emitted only in specific finite quanta, I have no objections to make; all of your applications also agree with this formulation. Now, this is not a characteristic of electromagnetic processes in vacuum but rather of the emitting or absorbing matter, and hence radiation does not consist of light quanta as it says in §6 of your first paper; rather, it is only when it is exchanging energy with matter that it behaves as if it consisted of them.[4]

Laue was apparently willing to concede that in the emission and absorption process light quanta were involved, but beyond that, he was adamant: light traveled through the vacuum of space as a wave, not as quanta. Laue was not alone in his belief. In 1905, the magnitude of Einstein's departure from the sanctioned belief about light was so unsettling that his particle theory of light was not accepted for two decades.

The Setting

Bertrand Russell is supposed to have asked this question: Is the world a bucket of molasses or a pail of sand? In less picturesque terms, the question becomes: Is the underlying reality of the world a seamless continuum or is it inherently grainy? Is it continuous or discontinuous? In mathematical terms the question is this: Is the world to be described geometrically as endless unbroken lines or is it to be counted with the algebra of discrete numbers? Which best describes Nature—geometry or algebra? This question has been described as "that same dilemma between atoms and the continuum which has given structure to the history of science since its opening in Greece."[5]

The contradictory ideas of continuity and discontinuity came together in Newton's physics. On the one hand, gravity fills all of space while, on the other hand, the sources of gravity are localized masses. Does continuous gravity spring from discontinuous masses? The same tension is found in James Clerk Maxwell's electromagnetism: "The energy in electromagnetic phenomena is [the same as] mechanical energy. The only question is, Where does it reside? On the old theories, it resides in the electrified bodies . . . On our theory it resides in the electromagnetic field, in the space surrounding the electrified and magnetic bodies."[6] Is electromagnetic energy continuous or discontinuous?

Discontinuity is frequently masked by apparent continuity. Dive from a pier into Lake Erie or from a diving board into a community pool and the water completely surrounds you; step out of the water and your skin is evenly covered by a layer of water. Plunge your hand into a pail of water and it is everywhere the same. Water appears to be continuous (as does Russell's bucket of molasses). We know, however, that water is not continuous. Start with a glass of water. Divide it in half. Divide it in half again. Keep dividing, again and again and again. As the water sample gets smaller, it becomes difficult to make a clean division, but there are precision instruments that allow the division process to continue until the basic entity of water is reached—one single molecule H_2O—and that ends it. No further division is possible. Even before we arrive at the last single molecule of water, however, the graininess of water becomes apparent. A few steps before the last division, a hundred or so individual water molecules are present, slightly separated from one another, moving relative to each other, striking and bouncing off neighbors. In between and surrounding the water molecules, there is nothing, just empty space. Water consists of particles.

Water and matter are discontinuous. Space, on the other hand, is continuous. All available evidence suggests that space extends unbroken in all directions from Earth and continues for distances that cannot be imagined. The space probe *Pioneer 10* was launched on March 2, 1972. It provided gorgeous pictures of Earth's planetary neighbors; however, after thirty years, it was no longer able to emit signals strong enough to be used here on Earth. Its last feeble signal was received on January 23, 2003. After that final signal, earthly contact with *Pioneer 10* was broken. Now this productive space probe moves in the dark of deep space like a lonely meteor. Moving inexorably along a straight line, it plays out, moment by moment, the concept of inertia as it heads toward the red star Aldebaran, the eye of Taurus, some sixty-eight light years away. Through the vastness of continuous space *Pioneer 10*, in essence a particle, moves toward its rendezvous with Aldebaran some 2 million years hence.

Is light continuous or discontinuous? Sunlight spreads over the front lawn uniformly illuminating every exposed blade of grass, dandelion, and grain of dirt. Light appears to be continuous. Not so, thought Isaac Newton in the seventeenth century. He regarded light as particulate. Newton, who thought in mechanical terms, noted that the behavior of light seemed best understood as streaming particles. Optical studies in the early nineteenth century, however, contradicted Newton's view. Thomas Young showed that two beams of light can join together to produce either a doubly bright beam or no beam at all. This evidence fit the interpretation of light as waves: when the two beams join crest on crest and trough on trough, constructive interference occurs and the resulting beam is doubly bright, but when two beams join crest on trough and trough on crest, destructive interference occurs and the two beams cancel each other out. (The type of interference just described is seen when two stones are thrown into a pond. As the ripples, originating from the impact sites of the two stones, cross each other, the crests and troughs are seen constructively and destructively interfering.) Interference is wave behavior; so is diffraction. Light, incident upon an opaque barrier, casts a shadow. A close examination, however, reveals that the light appears where it should not be: in the shadow zone. Just as sound waves can bend around corners, so do light waves. Diffraction, bending around edges, is wave behavior. Because of interference and diffraction, light was regarded as continuous.

In addition to optics, electricity and magnetism were also dominant fields of physical research in the nineteenth century. A great experimental physicist in England, Michael Faraday, originated the idea of continuous fields originating from electrical charges and magnets. Then in the 1860s, James Clerk Maxwell formalized the field concept in a synthesis that provided the basis for understanding all electromagnetic phenomena (known as Maxwell's equations). In addition, Maxwell's synthesis unexpectedly predicted the existence of a new kind of wave that was electromagnetic in nature. When the characteristics of these unknown electromagnetic waves were examined, it was discovered that they traveled at only one speed: 186,000 miles per second. Since that was precisely the speed of light, it was tempting to conclude that light waves were, in fact, electromagnetic. Finally, Heinrich Hertz experimentally confirmed the existence of electromagnetic waves and the conclusion was impossible to deny: light is an electromagnetic wave.

Once light was confirmed as an electromagnetic wave, a further problem had to be resolved; namely, what was the "stuff" or, as it was called, the medium, through which light moved as it propagated from star to star, from Sun to Earth, from streetlight to sidewalk? A wave, everyone presumed, had to have a medium to enable it to propagate. In a real sense, it is the medium that propagates a wave. The medium of choice for light was the theoretical ether: it had to fill all of space, exert no resistance to planets moving through it, and have very unusual properties that enabled it to propagate light at the tremendous speed of 186,000 miles per second. Everyone knew what light was, but without an experimentally validated medium, the picture was incomplete.

Determining the properties of the ether was an important issue in 1905. In spite of the uncertainties about its nature, however, the evidence and the considered judgments of physicists were overwhelming: light was a continuous wave. Against this evidence and judgment, Einstein proposed that light was discontinuous, that light consisted of particles. No other physicist, not one of his illustrious contemporaries, was thinking about light in this way. Einstein was alone. Although Einstein does not mention the ether in his March paper, he must have recognized that his particle theory of light negated the ether issue. Particles do not require a medium. Light, when considered as a stream of particles, does not require the ether. Only three months later, in his June paper, Einstein does away with the concept of ether. Did Einstein see his March paper as effectively preparing the way for the ether's demise?

Conflicting ideas are provocative. They stimulate thought and, as is the case with Newton's gravity, they are often linked. The space probe *Pioneer 10,* a discrete "particle," for example, moves through continuous space. In the process, continuity and discontinuity exist side by side. Sometimes conflicting ideas work in tandem. Sometimes they do not.

Consider, for example, a hunk of metal with a spherical cavity inside it. Drill a small hole from the external surface of the sphere to the cavity inside. Heat the metal until it is glowing. In the cavity, enclosed by the glowing metal, light bounces around and eventually comes streaming out through the hole. Examine the light. (This light is called blackbody radiation because the object, specifically, the hole leading to the cavity, behaves like a perfect absorber and a perfect radiator, which is called a blackbody.) When physicists examined the energy carried by this light, they could not explain it. Here we have light in direct contact with the atoms in the walls of the cavity. Here we have a vivid example of continuity (light) and discontinuity (atoms) side by side, and something is amiss. It took the revolutionary idea of the quantum, proposed by Max Planck in 1900, to begin the process of resolving the difficulty.

Planck introduced the quantum to explain the energy properties of light emitted by a blackbody. However, Planck's quantum idea did *not* apply to the light itself, but rather to the "resonators," as Planck called them, inside the cavity. It was the electrons in the walls of the cavity, oscillating at various frequencies, that emitted the light that eventually emerged from the hole. What Planck assumed was that the energies of the "resonators" were not continuous, but were broken into discrete units of energy, ε, proportional to the oscillator's frequency, ν, where $\varepsilon = h\nu$. This was the birth of Planck's constant, h. Since the constant h is so small,

$$h = 0.000000000000000000000000006626 \text{ erg-sec} = 6.626 \times 10^{-27} \text{ erg-sec},$$

the discrete bundles of energy are infinitesimal. Physicists "knew" that energy was continuous and when Planck broke energy up into discrete quanta, even though they were small, it was a major revolution. Einstein said later, "It was as if the ground had been pulled out from under one, with no firm foundation to be seen anywhere, upon which one could have built."[7]

Planck's seminal idea fell short. Planck's quantum was tethered to the "resonators" in the cavity; Einstein saw that the quantum had to be cut loose from its tethers. In his March paper, Einstein liberated the quantum.

The March 1905 Paper

The March paper exposes the way Einstein's mind worked. He begins his March paper in typical Einstein fashion. Here is the first sentence:

> There exists a profound formal difference between the theoretical conceptions physicists have formed about gases and other ponderable bodies, and Maxwell's theory of electromagnetic processes in so-called empty space.[8]

The issue appears here in stark contrast: the discontinuity of gases consisting of localized atoms and the continuity of light (electromagnetic processes) propagating in empty space.[9] Einstein immediately directs our attention to the fundamental nature of the issue. He often worked from generalizations or principles that were seen as contradictory—they fueled his imagination. Once he identified a contradiction, Einstein would generalize it and then be guided by its implications until a resolution was found, frequently in the form of profound new insights. In addition, he often showed how long-held preconceived notions were invalid. This same approach appears again in his May and June papers.

Is Einstein's "profound formal difference" merely an abstract, academic issue? Not at all. Many phenomena of Nature are a direct consequence of mutual interactions between unlocalized (continuous) radiation and localized (discontinuous) matter and it is when radiation and particles must be treated together, as in the case of blackbody radiation, that problems emerge. Forty years later, in his "Autobiographical Notes," Einstein referred to the 1905 period and wrote that "one is struck by the dualism which lies in the fact that the material point in Newton's sense and the field as continuum are used as elementary concepts side by side."[10] Einstein recognized this in 1905. He was convinced that there were problems with the juxtaposition of particles and waves. His March paper was a direct outcome of that knowledge.

Almost immediately in the March paper, Einstein acknowledged that the "wave theory of light" has been "excellently justified" for a range of optical phenomena. Specifically, Einstein meant phenomena like interference and diffraction (which beautifully supported the wave picture of light). However, these optical phenomena, for which the continuity of waves had proved so successful, were not concerned with either the emission ("creation" in his title) of light or the absorption (conversion) of light. Or, to express the problem as Einstein did in the opening of his paper, optical observations such as reflection and diffraction represent the behavior of light over an extended period of time whereas the emission or absorption of light by atomic matter occurs almost instantaneously. In this gentle fashion, Einstein prepares to show that when continuous light and grainy atoms are brought together, his "heuristic point of view" will prove worthy.

Indeed, it was only a heuristic view that Einstein presented. The March paper did not offer proof that light consisted of particles or quanta; rather, Einstein presented his particle idea as a provisional way of thinking about light, a way whose merit would be determined by its explanatory usefulness. Einstein's route to his objective is an odd one. At no point along the way can a reader anticipate how Einstein is going to bring off the claims with which he begins the paper. More than odd, however, the March paper is beautiful. It is a testimony to the power of Einstein's intuition.

To get at the problem, Einstein imagined a container with an inner cavity surrounded by reflecting walls. In the cavity are particles: gas molecules and electrons. In addition, there are electrons that are bound to the walls of the cavity in such a way that they can oscillate with a range of frequencies, like Planck's resonators, and, in so doing, the oscillators can absorb and emit electromagnetic radiation (light). Within the cavity we can imagine gaseous molecules and electrons flitting about, colliding with each other, colliding with the walls of the cavity, and with the oscillating electrons. At the same time, light is being emitted and absorbed by the oscillating electrons and is bouncing around in the cavity. The cavity and everything in it are at the temperature T.

Now, says Einstein, suppose the gas and the oscillating electrons, on the one hand, and the radiation and the oscillating electrons, on the other hand, are all in dynamic equilibrium so that each component—the gaseous atoms, the radiation, and the oscillating electrons—has the same average energy. In other words, no component loses or gains energy at the expense of another component. This seems all perfectly reasonable, but it does not work. Einstein shows that when radiation and the oscillating electrons are considered, a catastrophe is waiting to happen. If the oscillation frequency of the electrons is allowed to increase, the radiation energy increases without limit. The energy of the radiation becomes infinite. As infinite energy makes no physical sense, something was wrong.

In this imagined situation, Einstein brought light and particles together in a physically reasonable way and was led to a physically unreasonable result. For Einstein, the continuity of light and the discreteness of electrons and atoms were in conflict.

From an absurdity—light with infinite energy—Einstein moved quickly. Einstein's approach throughout this March paper was to swing between descriptions of atoms and radiation. The heart of this approach was based on the entropies of a sample of gas and a sample of light; specifically, how the entropy of samples of gas and light depends on their volumes. Entropy is the most fascinating and perhaps the most important concept of thermodynamics. The second law of thermodynamics is built around entropy. There are many ways to think about this concept (part of its mystique); one way is in terms of the order or disorder of a physical system. For a highly ordered state (a place for everything and everything in its place), the entropy is low; for a disordered state (everything thrown about in topsy-turvy fashion), the entropy is high. The second law of thermodynamics essentially says that all of Nature's processes that occur naturally, that is, occur without external intervention, are accompanied by an increase in entropy. Nature's way is to proceed from order to disorder or from low to high entropy.

The way the entropy of a sample of gas changes as its volume changes was well known. So with radiation Einstein began with the objective of discovering how the entropy of a sample of radiation depends on its volume; specifically, he set out to determine how its entropy changed when a sample of light was taken from a larger volume V_0 to a smaller volume V. To Einstein, the result was suggestive. Einstein found that for *both* a sample of continuous radiation and a sample of discrete gas the entropy's volume dependence was formally *the same*. Identical. Both for a gas and for radiation, when the volume is changed from V_0 to V, the entropy change is determined by the ratio V/V_0. Why should they be the same? Why should the entropy changes of a volume of gas, consisting of discrete atoms, and a volume of radiation, consisting of continuous waves, depend upon their volumes in *exactly* the same ways? However, with Einstein's goal in mind, we could ask why it should matter. A ratio of volumes does nothing, in and of itself, to adjudicate between continuity and discontinuity. This came next.

Having determined the entropies of a gas and light by the standard method, familiar to all physicists, Einstein next determined the entropies by a much newer method. As Einstein wrote, "The equation just found shall be interpreted in the following on the basis of the principle introduced into physics by Mr. Boltzmann, according to which the entropy of a system is a function of the probability of its state."[11] Thus, Einstein's route to his revolutionary result was to express the entropies of samples of a gas and radiation by means of the relatively new probabilistic ideas that came from the work of the Austrian physicist Ludwig Boltzmann.

The order-disorder idea associated with entropy can also be expressed in terms of probabilities. Order is associated with low probability; disorder with high probability. There are many, many different ways that items can be randomly thrown about a room to produce a disordered room or a sloppy room. With many ways to make a room sloppy, the most likely state of a room is sloppiness. Sloppiness is a high-probability state. By contrast, if there is a proper place for every item in a room, then out of all the many ways the items can be arranged, there is just one arrangement whereby each item is in its proper place. When each item is in its proper place, it is an ordered room; an ordered room is a low-probability state.

Boltzmann, who believed in atoms long before many of his contemporaries, explicitly expressed entropy in terms of order-disorder, that is, in terms of probabilities. Einstein asked exactly the same question: What is the change in entropy when a sample of gas, made up of randomly and independently moving atoms, is taken from a volume V_0 to a smaller volume V? To answer in Boltzmann's terms, Einstein had to ask: What is the probability of all the atoms of a gas, contained in the large volume V_0, suddenly finding themselves in a smaller portion of V_0 with the volume V? The answer is $(V/V_0)^N$, where the exponent N is the number of discrete, individual atoms in the sample. By means of the Boltzmann Principle, Einstein's result went from V/V_0 to $(V/V_0)^N$. When this latter result is brought into the expression for the change in entropy of a sample of gas as its volume changes from V_0 to V, it brings with it the exponent N, an inherent element of discreteness.

Up to this point, Einstein used information about the entropy of gases and radiation that was commonly known among scientists. Boltzmann's Principle was new, but it was known and available for Einstein's use. From this point on, however, Einstein was guided by his own intuition. With uncommon daring, he forged ahead.

First, Einstein tacitly assumed he could apply the same probabilistic approach to a sample of radiation as he had to a gas. Second, since the entropies of a sample of gas and a sample of radiation, determined by the older standard method, depended on their volumes in exactly the same ways, he assumed the entropies, determined by Boltzmann's newer probabilistic approach, would also depend on volumes in exactly the same way. So, with $(V/V_0)^N$ in his mind, he asked for radiation the same question he had asked for gases: If radiation, with a total energy E, is contained in a volume V_0, what is the probability that all the radiation energy will suddenly find itself contained in one small part of the volume of V_0, that is, in a smaller volume V? By analogy with his result for gases, Einstein intuits that the probability is $(V/V_0)^n$ where a new exponent n replaces the earlier exponent N. N is the number of discrete atoms; n, said Einstein, is the number of discrete light quanta. Each light quantum, continues Einstein, has an energy equal to $h\nu$. (Einstein did not use the symbol h, rather he used a collection of constants that equal h. I have used h here both for simplicity and because it is more familiar.) In terms of light quanta, the total energy of the sample of radiation is the energy of each quantum, $h\nu$, multiplied by the total number of quanta present, n, or $E = nh\nu$ where h is Planck's constant and ν is the frequency of the light.[12] Einstein then writes:

> From this we further conclude: . . . radiation . . . behaves thermodynamically as if it consisted of mutually independent energy quanta of magnitude $R\beta\nu/N$. [$R\beta\nu/N$ is equivalent to $h\nu$.][13]

Could Einstein conclude his March paper with this startling result? Einstein did not prove his hypothesis that light was particulate. And he knew this: the "as if" tacitly acknowledges the absence of proof. Certainly Einstein could not end the paper with this unfounded conclusion because he had not fulfilled the promise of his title: "a heuristic point of view." In fact, Einstein goes on to demonstrate that there are good reasons to accept his conclusion. He goes on to show that a particle view of light could resolve outstanding puzzles: Stokes's Rule, the photoelectric effect, and gaseous ionization.

Of the three applications that Einstein makes to demonstrate the "heuristic" value of his particle theory of light, it is the photoelectric effect that is most closely identified with the March paper. When light shines on a metal, electrons, called photoelectrons, can be ejected from its surface. Sometimes electrons are ejected; sometimes they are not. Experiments on the photoelectric effect produced some specific results, which were published in 1902. For example, bright light shining on a metal ejects a greater number of electrons than dim light, but surprisingly, the kinetic energy of the ejected electrons does not change as the light's brightness changes. Finally, for any particular metal, illuminated by a particular light source, the kinetic energies of the ejected electrons never exceed a particular maximum. These observations could not be explained by physicists adhering to the wave model for light.

Einstein's particle theory of light applied to the photoelectric effect was beautifully successful and stunningly simple. Of the three applications Einstein made, the photoelectric effect provided the most persuasive evidence for Einstein's particle view of light.

The photoelectric effect was discovered in 1887 by Heinrich Hertz, who was the same gentleman who proved that light was electromagnetic waves, ironically. But for the photoelectric effect, continuous waves cannot do the job. As Einstein wrote,

> The usual conception, that the energy of light is continuously distributed over the space through which it travels, meets with especially great difficulties when one attempts to explain photo-electric phenomena. . . . According to the view that the incident light consists of energy quanta, . . . the production of cathode rays [photoelectrons] by light can be conceived in the following way. The body's surface layer is penetrated by energy quanta whose energy is converted at least partially into kinetic energy of the electrons. The simplest conception is that a light quantum transfers its entire energy to a single electron; we will assume that this can occur.[14]

The wave theory of light fails the instant light strikes the surface of a metal. At its core, the problem is again one of continuity versus discontinuity. Experimentation later showed that, in the photoelectric effect, electrons are ejected from a metal's surface immediately. A light wave cannot do this. Simply put, a light wave, with its energy continuously distributed all along the wave, does not carry enough energy in that tiny portion of the wave that actually comes into contact with a particular electron on the metal's surface to kick it out of the metal instantly. Continuous energy incident on discontinuous electrons fails in the photoelectric effect. However, when the light energy is concentrated in a single particle—a single light quantum—that light quantum can collide directly with an electron, impart all or part of its energy directly to the electron, and immediately send it flying out of the metal. Discontinuous energy incident on discontinuous electrons succeeds.

Einstein's treatment of the photoelectric effect not only accounted for all the facts known about it, his treatment also predicted new facts. When light is considered as particles, each particle carrying an energy, $\varepsilon = h\nu$, then the kinetic energy of the ejected electrons should increase directly as the frequency of the incident light increases; there should exist a cutoff frequency, that is, a minimum frequency below which the light is unable to eject electrons.

When Einstein completed the formal part of his paper and concluded that light behaves "as if" it consisted of quanta, he had not achieved his objective; however, with the application of his particle view of light to the photoelectric effect, Einstein's light particles acquired a heuristic advantage.

The Response

Einstein's application of the photoelectric effect was more than successful. He anticipated qualitative and quantitative features of the photoelectric effect whose details became known only later. For example, the energy of Einstein's light particles increase with the frequency of the light: high-frequency violet light particles (large ν) carry more energy than low-frequency red light particles (small ν). This means that as the color changes from ultraviolet to violet to green to red to infrared (decreasing frequency), at some point there is a cutoff when the energy of the light particle becomes too small to eject an electron. The "cutoff frequency" was eventually verified by experiment. The speed with which the electron leaves the metal should increase in direct proportion to the light particle's energy, that is, to its frequency. It had been qualitatively demonstrated that electron speeds varied with different light sources, but eventually Einstein's precise predictions were quantitatively verified. In fact, nothing in Einstein's March paper had to be altered.

The application of a particle model of light to the photoelectric effect was indeed, as stated, both successful and simple. Today, the photoelectric effect compellingly makes the case for a particle view of light. But not in 1905. Physicists then, and for years to come, refused to budge from their commitment to a wave theory of light.

On July 6, 1907, Max Planck wrote a letter to Einstein:

> I look for the meaning of the elementary quanta of action [light quanta] not in a vacuum [the domain of continuous light waves], but at points of absorption and emission, and I believe that the processes in the vacuum are *accurately* described by the Maxwellian equations. At least, I don't as yet see any compelling reason for departing from this assumption which, for the moment, seems to me the simplest one, and one which characteristically expresses the contrast between ether and matter.[15]

Or listen to Planck six years later in 1913, during the speech in which he nominated Einstein for membership into the Prussian Academy of Sciences in Berlin: "That sometimes, as for instance in his hypothesis on light quanta, he may have gone overboard in his speculations should not be held against him too much, for without occasional venture or risk no genuine innovation can be accomplished even in the exact sciences."[16] Eight years after the publication of Einstein's "heuristic viewpoint," Planck was still rejecting it.

An even more dramatic rejection came in 1916. In a series of experiments, Robert A. Millikan confirmed in definitive fashion all the predicted properties of the photoelectric effect that came out of Einstein's March paper. There was a cutoff frequency; the

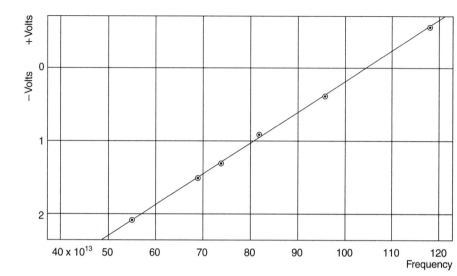

The straight line (with a slope of *h*) is predicted by Einstein's theory. The encircled points are data points as measured by Robert Millikan. The experimental data points fall directly along the line, consistent with Einstein's theory.

kinetic energy of the ejected electrons did increase directly as the frequency of the incident light increased. Was Millikan pleased? No, he was chagrined. Eleven years after Einstein's 1905 March paper, Millikan published his experimental results confirming Einstein's predictions. He wrote: "We are confronted, however, by the astonishing situation that these facts were correctly and exactly predicted nine years ago by a form of quantum theory which has now been pretty much abandoned."[17] In this same paper, Millikan also characterized Einstein's paper as a "bold, not to say reckless, hypothesis of an electro-magnetic light corpuscle of energy *hv*, which . . . flies in the face of thoroughly established facts of interference."[18] A year later, after again acknowledging that "the equation of Einstein seems to us to predict accurately all the facts which have been observed,"[19] Millikan goes on to write:

> Despite then the apparently complete success of the Einstein equation [for the photoelectric effect], the physical theory on which it was designed to be the symbolic expression is found so untenable that Einstein himself, I believe, no longer holds to it, and we are in a position of having built a very perfect structure and then knocked out entirely the underpinning without causing the building to fall. It stands complete and apparently well tested, but without any visible means of support. These supports must obviously exist, and the most fascinating problem of modern physics is to find them. Experiment has outrun theory, or, better, guided by erroneous theory, it has discovered relationships which seem to be of the greatest interest and importance, but the reasons for them are as yet not at all understood.[20]

Millikan was wrong in rejecting Einstein's corpuscular theory of light, and he was also wrong about Einstein. One year before Millikan's claim that Einstein had abandoned the quantum theory of light, Einstein wrote a letter to Michele Besso in which he said that the existence of "the light quanta is practically certain."[21]

Einstein's confidence was not bolstered by the support of other physicists. Even Niels Bohr, one of the strong advocates of the quantum, spoke out against light quanta in his 1922 Nobel Prize address: "The hypothesis of light-quanta . . . is not able to throw light on the nature of radiation."[22] Compare Bohr's statement to Einstein's early anticipation of wave-particle duality in 1909:

> I have already attempted earlier to show that our current foundations of the radiation theory have to be abandoned . . . It is my opinion that the next phase in the development of theoretical physics will bring us a theory of light that can be interpreted as a

kind of fusion of the wave and the [particle] theory . . . [The] wave structure and [the] quantum structure . . . are not to be considered incompatible.[23]

It was not until 1923, through the work of Arthur Compton at Washington University in St. Louis, that Einstein's light-particle idea was made undeniably plain to physicists. In his telling experiments, Compton set out to study how X-rays and gamma rays are scattered by matter.[24] Gamma rays and X-rays are very high frequency light and, if Einstein is to be believed, very high energy X-ray particles. Compton discovered that when X-rays "hit" an electron, a collision occurs that is exactly like the collision between two billiard balls. Compton's experiment provided compelling evidence for physicists and, after seventeen years, Einstein's basic light-quantum idea became respectable.

Contemporary physicists regard the photoelectric effect as such a vivid testimony to the particle nature of light that they typically refer to Einstein's March paper as "the photoelectric effect paper." They could refer to it just as accurately as "the Stokes's Rule" paper. As is frequently the case, however, the history invoked by physicists is sometimes misleading, and this is an example. The March paper was the "particle theory of light" paper. The significance of the particle theory of light goes far, far beyond the photoelectric effect—the photoelectric effect merely provided evidence in favor of it.

In 1926, one year before the theory of quantum mechanics was completed, Gilbert N. Lewis introduced the term "photon," which subsequently became the name for Einstein's light particle. With the completion of quantum mechanics, the photon soon became indispensable. In the quantum theory of fields and in quantum electrodynamics (QED), the photon is the means by which electric charges interact with each other; in short, the photon is the mediator of electromagnetic forces. One century after Einstein made the audacious suggestion that light is a particle, the photon plays many roles in physics. Today, Einstein's light particle is indispensable.

Einstein submitted his March paper to the editors of *Annalen der Physik* on the seventeenth of that month. Near the end of his life, he was still thinking about light quanta. In a letter to Besso, Einstein acknowledged that after fifty long years of "conscious brooding" over the question "What are light quanta?" he was no closer to the answer.[25]

A New Determination
of Molecular Dimensions

A. Einstein

Einstein's doctoral dissertation was completed on April 30, 1905, but not published until 1906. It was published in *Annalen der Physik*, volume 19, pages 289–305.

Albert Einstein in 1929, the year he received the Planck Medal.

Molecular Dimensions

To appreciate Einstein's doctoral dissertation, it must be examined in the context of its era and in the context of Einstein's May paper. The famous May paper is always included in the lineup of great papers that seemed to pour out of Einstein during his incredible year. Einstein's April paper, however, is largely ignored. This is an unfortunate oversight for several reasons. First, Einstein's April paper fed directly into his May paper. Second, his dissertation reveals an aspect of Einstein's mind not apparent in his other papers; it shows a different side of Einstein. Finally, there is a human-interest angle to Einstein's dissertation: Einstein's road to a dissertation acceptable to the authorities almost ended in failure.

Einstein completed his dissertation on April 30, but he did not submit it to the University of Zurich for almost three months. There is a possible explanation for this three-month delay. It may well be that in April and May, Einstein had abandoned his hope of getting a doctorate and considered his April paper to be just another paper. In fact, two years earlier, Einstein had essentially given up his quest for an advanced degree. On January 22, 1903, he wrote to his friend Michele Besso, "I will not go for a doctorate, because it would be of little help to me, and the whole comedy has become boring."[1]

Was Einstein's pursuit of a doctorate really a comedy? Only if comedy becomes the means of coping with sustained travail and disappointment. In October 1900, Einstein began a dissertation in Professor Heinrich Weber's laboratory at the Zurich Polytechnic on the thermoelectric Thompson effect. For reasons unknown, this research effort did not work out, and his first dissertation attempt was a failure. After this, and with no apparent supervision, he began work on a second dissertation, this time on attractive forces between molecules in liquids, and in November 1901, he submitted it to Professor Alfred Kleiner at the University of Zurich. He wrote to his future wife Mileva Marić on November 28 "that he won't dare reject my dissertation."[2] His dissertation was rejected nevertheless and three months later the fees associated with the dissertation procedure were returned.

Einstein's 1903 boredom was not permanent; however, there were two more acts in Einstein's dissertation comedy. In the summer of 1905, Einstein once again set the doctoral wheels in motion by submitting a dissertation. According to his sister, Maja Einstein, he first submitted his recently completed paper on the theory of relativity (the June paper). It was rejected. Maja suggested that her brother's third failure was because the content of the relativity paper "seemed a little uncanny to the decision-making professors."[3] With this third rejection the comedy nears an end.

In the summer of 1905, Einstein had an abundance of potential dissertation candidates: the March, April, May, and June papers. With the June paper rejected, Einstein picked his earlier April paper, which, he thought, was void of any novel or startling ideas that could offend his professors. In addition to the strange content of the June relativity paper, it was entirely theoretical. In 1905, theoretical physics was still something of a novelty. Many physicists still thought that all physics should be experimental. So Einstein made a strategic choice: the April paper was not speculative, and it was directly connected to experiment. He again submitted a dissertation to Alfred Kleiner of the Zurich faculty on July 20, 1905. Kleiner circulated Einstein's dissertation among the members of the physics faculty along with his comments: "The arguments and calculations are among the most difficult in hydrodynamics and could be approached only by someone who possesses understanding and talent for the treatment of mathematical and physical problems, and it seems to me that Herr Einstein has provided evidence that he is capable of occupying himself successfully with scientific problems."[4] At this point in Einstein's dissertation saga, the last act of the comedy occurs. Either before or after Kleiner circulated Einstein's dissertation, he returned it to Einstein because it was deemed too short. According to Einstein, he added one sentence, after which it was immediately accepted. (There is some uncertainty about this last step in the Einstein dissertation story: Einstein apparently told this story to his biographer Carl Seelig.) Einstein's laughter must have rung out when the curtain fell on the five-year comedy.

In the overall context of Einstein's work, his dissertation stands apart. Einstein's professional papers are characterized by his penetrating insights into foundational issues, his arrival at those insights by a clever route, and his expression of those insights in formal ways that were often simple. Einstein's papers were often based on idealized physical systems and did not lead quickly to practical applications. With a few exceptions, Einstein's papers were not highly mathematical. By contrast, in his dissertation he focused on a liquid solution that required considerable mathematics and, as Kleiner implied, involved "difficult" calculations. Whereas the April paper did demonstrate Einstein's keen insight and while there were certainly clever aspects to it, his approach was vigorous, at times even forceful. Finally, his dissertation stands apart because it stimulated numerous practical applications, which explains why it is one of Einstein's most cited papers.

His dissertation also prepared the way for his May paper. The basic concepts Einstein employed in April were utilized again in

May. Just ten days after he finished his April paper, he submitted his May paper on Brownian motion for publication.

The Setting

The idea that all material is made up of atoms seems incontrovertibly ancient. It is true that atomism, the idea that matter consists of irreducible entities called atoms, goes back to the ancient Greeks. Indeed, ever since Leucippus and his student Democritus proposed the idea of atoms in the fifth century B.C., atoms have hovered in the minds of scientists—sometimes center stage, sometimes in the wings. However, the idea of atoms is one thing; an empirical, scientifically compelling case for atoms is quite another. In the early nineteenth century, chemists were successful in establishing laws by which the chemical elements combined to form molecules. The work of John Dalton, for example, strongly implied the existence of atoms. However, such inferential evidence was not sufficient cause for some scientists—chemists and physicists—to adopt an atomic view of matter. It was not until the early years of the twentieth century that an atomic theory was realized based on experimental evidence tied together by coherent theoretical constructs. In 1905, in spite of the discoveries of radioactivity and the electron, there remained a few reputable scientists who still regarded the evidence for atoms as insufficient to justify their belief.

We might suppose that the anti-atomists were second-rate scientists. Not so. In fact, those who did not accept the concept of atoms formed a small but impressive group that included, among others, the chemists Marcellin Berthelot (1827–1907), Wilhelm Ostwald (1853–1932, Nobel Prize 1909), Jocobus Hendricus Van't Hoff (1852–1911, Nobel Prize 1901), the physicist Ernst Mach (1838–1916), and the mathematician and physicist Henri Poincaré (1854–1912). Some were strongly opposed, others less so. The issue for them was twofold. First, there was a paucity of evidence that the skeptics regarded as an acceptable substitute for the direct observation of atoms. Second, there were strongly held philosophical views as to what constituted proper science. The anti-atomists' position was that science should be based exclusively on concrete empirical facts; conceptual ideas and hypotheses, they thought, corrupted science and therefore should be eliminated from the scientific canon. In principle, they thought, atoms could never be seen and therefore would always remain a hypothetical construct. Therefore, atoms were unacceptable.

Einstein could not see atoms, but he believed in them. In this belief, Einstein implicitly took issue with the philosophical position of those like Ernst Mach who rejected the atomic theory of matter. Later, in his "Autobiographical Notes," Einstein described the influence of philosophical attitudes on science:

> The antipathy of those scientists toward the atomic theory can indubitably be traced back to their positivistic philosophical attitude. This is an interesting example of the fact that even scholars of audacious spirit and fine instinct can be obstructed in the interpretation of facts by philosophical prejudices. The prejudice—which has by no means died out in the meantime—consists in the faith that facts by themselves can and should yield scientific knowledge without free conceptual construction.[5]

Einstein rejected this exclusively fact-based philosophy. Such a philosophy constrained the creative mind which, he believed, was capable of developing general laws that went beyond the description of known facts and, in the process, led to new knowledge.

For some time Einstein had been thinking about atoms, thinking about how to demonstrate their existence in a convincing fashion. In a March 17, 1903, letter to Michele Besso, Einstein asked this question: "Have you already calculated the absolute size of the ions under the assumption that they are spheres and that they are large enough to permit the application of the equations of the hydrodynamics of viscous liquids?"[6] In this letter, Einstein described exactly what he would do two years later in his dissertation.

The subject of Einstein's doctoral dissertation confirmed his belief in the atomic theory. Even more, his dissertation demonstrated his desire to provide new evidence that would further strengthen the atomic view. When Einstein proposed a particle nature of light in his March paper, he put himself in opposition to virtually the entire scientific community. There was solid unanimity about the nature of light and Einstein's quantum view of light was almost unanimously rejected. Einstein's support of the atomic hypothesis, however, did not put him in a stand-alone position. Most scientists by 1905 had accepted atoms as the basis of the material world. Only a few skeptics remained. It was Einstein's dissertation and its sequel, the May paper, that stilled the voices of the atomic skeptics.

The April 1905 Paper

By comparison with Einstein's other 1905 papers, the April paper has its feet solidly on the ground. There are no allusions to grand themes such as continuity versus discontinuity. There is no contradiction identified that challenged existing physical theory. There are no allusions to fundamental misconceptions.

The first sentence of the April paper read as follows:

> The earliest determinations of the real sizes of molecules were made possible by the kinetic theory of gases, whereas the physical phenomena observed in liquids have thus far not served for the determination of molecular sizes.[7]

Using gaseous samples, atomic diameters had been determined to be in the range of 1 to 4 Angstroms (1 Angstrom = 1 × 10^{-8} centimeter [cm]). Since several atoms come together to form molecules, their dimensions would be somewhat larger. Einstein sought to determine molecular size in liquids.

In the first paragraph of his paper, Einstein describes his strategy:

> It will be shown in this paper that the size of molecules of substances dissolved in an undissociated dilute solution can be obtained from the internal friction [viscosity] of the solution and the pure solvent and from the diffusion [rate] of the dissolved substance within the solvent.[8]

The key words are viscosity and diffusion. All liquids (and gases) possess a property called viscosity that is expressed in terms of a coefficient of viscosity. A liquid's viscosity is the measure of the resistance an object confronts when moving through it. Molasses is more viscous than water. Diffusion occurs when there is an uneven concentration of a dissolved substance in a liquid (or gas). The dissolved substance diffuses from a region of high concentration to a region with lower concentration.

Liquids are inherently messy; they are complicated and, without simplifying assumptions, they defy quantitative analysis. But it was to a liquid that Einstein directed his attention for his dissertation topic. More specifically, Einstein considered a solution in which a material was dissolved in a pure liquid. Because of data available to him, Einstein applied the results of his analysis to a solution of sugar dissolved in water. The question he wanted to answer was, "What is the dimension of the sugar molecule?"

Einstein first considered viscosity. He pictures the sugar molecules as little round balls. Each sugar ball is surrounded by water and these imaginary balls travel through the water. But the sugar molecules do not glide through the water freely. Just as friction impedes the motion of a book sliding across the floor, the viscosity of water impedes the motion of a sugar molecule.

Einstein broke his problem into two parts. First, in his imagination, he considered the flow of molecules in a pure liquid. If the static pressure of the liquid increases, say, from right to left, the liquid will move toward the low-pressure region, from left to right. This left-to-right motion of the liquid molecules will be opposed by the viscosity of the liquid. A liquid's viscosity can be measured and the result of the measurement is called the liquid's coefficient of viscosity.

Next, Einstein imagined the addition of sugar to the water. Of course, the sugar would dissolve in the water and its addition would change the viscosity from that of pure water. Einstein made a number of assumptions to simplify the analysis. For example, he assumed that relatively few sugar molecules were dissolved in the water; in other words, it was a dilute solution. He assumed that the sugar molecules could be treated like spheres that were larger than the surrounding molecules of the liquid. There is a bit of irony here: his objective was to determine the size of the discrete sugar molecules, but in treating the water molecules as small in comparison to the sugar molecules, he effectively treated the surrounding water as a continuous medium—discontinuity in continuity.

With these simplifying assumptions guiding his thoughts, Einstein considered the flow of the sugar molecules, assumed to be spherical, through the surrounding liquid. Once again, the spheres will move through the liquid in response to a pressure variation and that motion will be opposed by the viscosity. However, as stated, the viscosity with the spheres dissolved in the liquid is different from that of the pure liquid. Einstein was able to derive a relationship between the two viscosities in terms of the total volume occupied by the dissolved molecules.

The relationship Einstein derived between the two viscosities contained two unknowns. The two viscosities themselves could be measured. The unknowns in Einstein's relationship were Avogadro's number, which was not known very accurately in 1905, and the radius of the dissolved spheres (the sugar molecules), which was what Einstein wanted to know.

At this point, Einstein had one relationship with two measurable viscosities and two unknowns. As students of algebra know, one equation with two unknowns cannot make known the unknowns. A second equation, with the same two unknowns, was needed.

After viscosity, Einstein considered diffusion. First, I'll provide an example of diffusion. If a bottle of strongly aromatic perfume is opened in a small room, the smell of the perfume is soon noticeable throughout the room. This is because of diffusion. The perfume molecules are most concentrated above the bottle and the perfume molecules diffuse from the bottle (high concentration) to other parts of the room (lower concentration). The diffusion process continues until the concentration of perfume molecules is uniform throughout the room.

Now back to the April paper. Once again consider a dilute solution of sugar dissolved in water. The sugar molecules are large in comparison to the water molecules. Einstein imagined a force acting on the sugar molecules, a force acting toward the left. As a result of the leftward acting force, the sugar molecules will move toward the left and, as they do so, the concentration of sugar molecules will increase on the left relative to the right. With the sugar concentration increasing from right to left, the diffusion of sugar molecules will take place left to right, that is, from high to low concentrations. One can imagine an equilibrium situation in which the number of sugar molecules moving to the left (because of the applied force) equals the number of sugar molecules moving to the right (because of diffusion). From this equilibrium condition, Einstein was able to determine the diffusion coefficient for sugar molecules in water.

The diffusion coefficient for a sugar solution can be measured.

Einstein's expression for the diffusion coefficient again contained two unknowns: Avogadro's number and the radius of the dissolved sugar molecules. Since both the coefficient of viscosity and the diffusion coefficient could be determined experimentally, Einstein had two equations and two unknowns that he easily solved. He determined that N = Avogadro's number = 2.1×10^{23} and that the radius of the dissolved sugar molecule was 9.9×10^{-8} cm.

Avogadro's number was not known with any great accuracy in 1905, but all methods for determining this important number hovered around 10^{23}, so Einstein's theoretical value was very reasonable. Also, his result for the size of sugar molecules was in agreement with other estimates of molecular size. On April 30, 1905, Einstein had every reason to be confident in his results.

The Response

Einstein's dissertation was accepted in the summer of 1905 and with that, Mr. Einstein became Dr. Einstein. The story of Einstein's dissertation, however, was not over. When his dissertation was accepted by the professors at the University of Zurich, Einstein sent it to the editor of *Annalen der Physik* for publication. However, it was not published until 1906. The editor, Paul Drude, knew of some viscosity and diffusion data that were more accurate than the data Einstein had used in his dissertation. Using his editorial prerogative, Drude asked Einstein to make use of the improved data. Einstein did this in an addendum to his original manuscript and with that, his paper was published.

In 1909, a French physicist, Jean Perrin, was working on the implications of Einstein's May paper on Brownian motion and wrote to Einstein with a question. Einstein referred him to his dissertation, the April paper, which was available in the journal *Annalen der Physik*. Perrin put a student, Jacques Bancelin, to work on an experiment that was based on Einstein's dissertation. Instead of using sugar, however, Bancelin suspended precisely prepared microscopic mastic globules with known radii in the water. When Bancelin compared his results with Einstein's theory, there was a discrepancy. In response, Einstein repeated his calculations, could find no error, and was unable to resolve the discrepancy. He asked a Zurich physicist, Ludwig Hopf, to look at his calculations to see if there were any errors. Hopf found a mistake, which Einstein corrected and sent to the editor of *Annalen der Physik* with the experimental results obtained by Bancelin. In the end, the theory presented in Einstein's dissertation and the best experimental data available were in beautiful agreement. And with that, his dissertation finally was behind him.

Einstein's dissertation provided a way to determine molecular dimensions in a liquid. This did not, in and of itself, enable the atomic skeptics to see or to touch an atom or a molecule. However, it did provide additional evidence, this time evidence from liquids, for the atomic theory. More significant, Einstein's May paper on Brownian motion is a direct descendent of his dissertation and, as we shall see, when the April and May papers are considered together, the case for atoms becomes convincing.

As stated earlier, Einstein's April paper had a more applied character than did his other 1905 papers. Because of this, the theory laid out in his dissertation has been applied in many practical ways ranging from analysis of the motion of sand particles in cement mixes to the motion of casein micelles in cow's milk and the motion of aerosol particles in clouds.[9] These many applications mean that Einstein's April paper is cited often. One indicator of the influence of a scientific paper is how frequently other scientists refer to it in their own papers. Which papers written before 1911 from all the sciences are most frequently cited fifty years later? For the fourteen-year period from 1961 to 1975, the eleven most frequently cited papers written before 1911 include four papers written by Einstein. His April paper was third on the list. No other scientist had two papers on the list. The fourth paper on the list is Einstein's May paper on Brownian motion. The other two Einstein papers on the list are a 1911 paper (number five) and a 1910 paper (number eleven).[10]

This list is surprising but also misleading. It is surprising because Einstein's April and May papers are certainly not as well known as his June paper on the theory of relativity. And the March paper was certainly more revolutionary than either his April or May paper. Why aren't the March and June papers on the list? Einstein's quantum paper and his relativity paper undergird much of modern physics. As such, they are so important that they are taken for granted. When a paper is so important that it could be cited in almost every paper, it is cited in almost no paper. That explains why Einstein's March and June papers do not appear on the citation list.

Einstein's April paper, which by default became his dissertation, does not rank with either the March or the June paper. But as the basis for the May paper on Brownian motion, Einstein's April paper stands comfortably in the company of the other 1905 papers.

On the Movement of Small Particles Suspended in Stationary Liquids Required by the Molecular-Kinetic Theory of Heat

A. Einstein

The paper was received by the editor of *Annalen der Physik* on May 11, 1905, and published that same year in *Annalen der Physik*, volume 17, pages 549–560.

Albert Einstein in 1932, the year he was appointed a professor at the Institute for Advanced Study in Princeton, N.J. His original plan was to divide his time between Princeton and Berlin. However, one year later, he left Germany for the United States, arriving in Princeton on October 17, 1933. Einstein lived in the United States for almost twenty-two years, but in many ways he remained a European.

"Seeing" Atoms

Einstein begins his May paper with this sentence:

> It will be shown in this paper that, according to the molecular-kinetic theory of heat, bodies of microscopically visible size suspended in liquids must, as a result of thermal molecular motions, perform motions of such magnitude that these motions can easily be detected by a microscope.[1]

When words are composed for publication in a scientific journal, they are typically written in a formal, bland, and often boring style. Einstein's papers were formal, perhaps bland, but they were certainly not boring. Behind Einstein's formal language in the first sentence of his May paper was a ferocious mental image based on a supposition he took for granted, but, in the strictest sense, remained a supposition. The supposition was that all matter consisted of individual atoms or molecules. Einstein believed this. The mental image was of a serene liquid filling a container; however, Einstein's attention was not on its external serenity, but on its internal chaos. He saw a churning cauldron of molecules moving rapidly in all directions, banging into each other and crashing into the container's walls. Most important, pollen particles, visible through a microscope, were suspended in the liquid, and the unseen liquid molecules mercilessly bombarded the intruding pollen particles and set them in motion along random zig-zagging paths called Brownian motion. It was a compelling image.

In the second paragraph of this paper, Einstein lays bare the issues:

> If it is really possible to observe the motion to be discussed here, along with the laws it is expected to obey, then classical thermodynamics can no longer be viewed as strictly valid even for microscopically distinguishable spaces, and an exact determination of the real sizes of atoms becomes possible. Conversely, if the prediction of this motion were to be proved wrong, this fact would provide a weighty argument against the molecular-kinetic conception of heat.[2]

Once again, the issues were profound. In the March paper Einstein tackled continuity versus discontinuity. In the May paper it is atoms versus thermodynamics. On the one hand, if Brownian motion *is* observed, as Einstein, by means of this May paper, tacitly assumes it will be, then the validity of thermodynamics is limited. Thermodynamics, the physical theory that is generally concerned with all thermal phenomena, is one of the great theories of physics. The three laws of thermodynamics were solidly supported by experiment. To limit the absolute validity of the established theory of thermodynamics has major implications. On the other hand, if Brownian motion is *not* observed, then the existence of atoms is in grave doubt. Did matter consist of atoms and molecules, or did it not? Whatever the outcome, the consequences of Einstein's May paper carried enormous significance for physics. Einstein had a nose for the big issues, and the stakes were high.

The Setting

Einstein's professional career advanced in stages. He was about halfway through the first stage of his career in 1905, still outside the community of physics working at the patent office in Bern,

Switzerland. In short, the patent office was not an environment conducive to scientific debate. Also, Einstein was only twenty-six years old in 1905 and largely unknown, so he did not receive letters from other well-established physicists who might have alerted him to important new developments. Finally, add to this mix Einstein's penchant for working alone and it becomes clear why his publications typically contained so few references to the scientific literature. These facts also explain why Einstein seemed unaware of things he might have been expected to know.

Einstein's May paper demonstrates his detachment from the active science community. In this paper Einstein cites just one reference. Although he explicitly acknowledged that he was aware of the existence of Brownian motion, he also confessed he had little information about it. For Einstein, the May paper was about "the movement of small particles suspended in stationary liquids," which, he stated, might be Brownian motion. It was only later that Einstein's May paper came to be called the "Brownian motion" paper.

Brownian motion is named for the English botanist Robert Brown. This phenomenon, which was observed by others before Brown, is the zig-zagging random motion of tiny particles, such as pollen particles, that is observed when the particles are suspended in a liquid. The zig-zag motion typically is not seen with the unaided eye; a microscope is required. In 1828 Brown showed that these zig-zagging particles were not living organisms. Prior to Brown's experiments, these moving particles were thought to be some kind of life form, "animalcules," propelling themselves through the liquid.[3] Brown sprinkled finely ground inorganic matter like finely ground glass, metals, and rocks into a liquid and the same herky-jerky motions of the inorganic particles were observed. If Brownian motion was not biological in origin, then the causes were clearly physical and a physical explanation was called for.

Why did Einstein choose to analyze the motion of small particles suspended in a liquid? To begin, Einstein believed in atoms and he saw in this peculiar motion of tiny particles a way to convince those scientists who still rejected the idea of atoms. But also his approach to this phenomenon was a logical extension of both his April paper and his earlier work on statistical fluctuations, as well as his earlier work that implicitly raised fundamental questions about the nature of physical reality.

According to the atomic theory, the atoms in a gas or liquid are in random, ceaseless motion. (In solids, atoms are located at particular sites and their motions are constrained to the immediate region around their particular sites.) The observed properties of a gas or liquid are a product of the unseen randomness in the behavior of its constituent atoms. More accurately, the observed properties of a gas are determined by *the average behavior* of its constituent atoms. However, in any random system, fluctuations can occur during which the random elements making up the larger system depart from their average behavior. Earth's atmosphere is a system whose behavior is driven by a number of influences that are inherently random in nature—including the random motion of the molecules making up the atmosphere. For example, an unusual number of molecules in the atmosphere can move into a localized region, which results in a high-pressure region. Of course, the opposite can occur and low-pressure regions result. High-pressure and low-pressure regions come and go. When the atmosphere behaves in an average way, average weather patterns are observed. When an extended period of unusual weather occurs, a series of days with temperatures well above or below average, a fluctuation is responsible.

Einstein recognized that since randomness was an inherent consequence of the atomic theory of matter, fluctuations had to be an inherent part of atomism. The basic macroscopic properties of a gas that we measure, properties like the temperature of the gas, the pressure of the gas, and the volume of the gas, are the properties that essentially define the sample of gas. For example, a defining property of a gas is its temperature, which is determined by the average speed of the unseen molecules making up the gas. Another defining property of a gas is its pressure, which is determined by molecules colliding with the walls of the container: the more often and the harder they bang into the wall, the higher the pressure. As long as the atoms behave in their average way, then the temperature and pressure remain reliable, stable properties of the gas. However, if there are fluctuations during which atoms deviate from their normal behavior and if the fluctuations are large, then it raises questions about the stability of measured properties and, in the process, raises basic questions about classical thermodynamics.

Consider Newtonian gravitation. Newton's gravitational force law describes the attractive force that holds Earth to the Sun. This force depends on three fixed parameters: the mass of the Sun, the mass of Earth, and the distance separating them. This force is independent of what the molecules making up Earth or the Sun are doing. No fluctuations compromise this law. Newton's gravitational force law is stable. By contrast, the laws of thermodynamics apply to material systems with their underlying atomic randomness. According to the second law of thermodynamics, for example, the entropy of a material system will always increase. But fluctuations can occur within localized regions of a system and, in that region, the entropy could temporarily decrease. This raises a question: Are the laws of thermodynamics generally true or, as Ludwig Boltzmann argued, are they true only in a statistical sense? This was the issue Einstein referred to at the beginning of the paper.

As in all statistics, the larger the number being averaged, the more stable is the average. In any sample of a gas or liquid, the number of molecules is so enormous that departures from average behavior are not to be expected. In fact, the leading authorities in statistical physics, Ludwig Boltzmann and J. Willard Gibbs, maintained that departures from average behavior would never be observed. This is essentially equivalent to saying that statistical fluctuations cannot be observed. Einstein did not accept this conclusion and sought the physical means to demonstrate and to observe statistical fluctuations.

Einstein believed the zig-zagging motion of tiny particles suspended in a liquid was the result of statistical fluctuations in the motions of the liquid's molecules. So "Brownian motion" was the physical means Einstein sought; even more, it was a means to answer not one, but three basic questions of physics. Atoms, yes or no? Statistical fluctuations, yes or no? The laws of thermodynamics: are they absolute or statistical? The answers to these three questions were conditioned on four "ifs:" If Einstein could successfully develop a theory to describe the motion of particles suspended in a liquid, if that theory was based on the statistical fluctuations in the random motions of liquid molecules, if the theory was experimentally testable, and finally, if experimental results were in accord with Einstein's theory, then the three answers would be: Atoms? Yes. Statistical fluctuations? Yes. And the laws of thermodynamics? Statistical.

One of the most provocative elements in the setting for Einstein's May paper was the kinetic theory of heat. The image of a gas provided by the kinetic theory made gases an attractive target for analysis. The image of a liquid provided by the kinetic theory is very much the same as for gases, that is, molecules of the liquid moving about randomly, colliding into each other and with anything else that gets in their way. However, the molecules of a liquid are, on average, much closer to each other. Because of their closeness, they interact with other nearby liquid molecules. This interaction makes liquids a more challenging system to analyze and understand. The image of a liquid with its random molecular motion invited speculation about Brownian motion. Some scientists envisioned water molecules banging into a suspended pollen particle and moving it; other scientists regarded the suspended particle as too massive to be influenced by a molecular collision.

The setting for Einstein's May paper was profoundly substantial: atoms, statistical fluctuations, and classical thermodynamics hung in the balance. The setting was timely: his dissertation prepared the way. With Einstein able to draw material from the April paper for the May paper, a mere ten days elapsed between the time he finished the former and the time he submitted the latter for publication. The setting was challenging: establishing the reality of atoms and getting them accepted into the canon of science as the building blocks of matter. And the setting was inviting: the image of a liquid seething with molecular motion called out to be examined as the possible cause of Brownian motion.

The May 1905 Paper

The mental image of fast-moving molecules bombarding a large suspended particle is intriguing. But there are problems. Einstein assumed that suspended particles, such as grains of pollen, had a diameter of 1/1,000 millimeter (mm), that is, 0.0001 centimeters (cm). In a molecular environment, Einstein's particles are gigantic. If we assume that a water molecule is a sphere (actually, a rather bad assumption), its approximate diameter would be about 1×10^{-8} cm, which is much, much smaller than a suspended particle such as Einstein considered. In fact, a water molecule striking a suspended pollen particle is like a baseball with a radius of 2.94 inches (7.47 cm) hitting a sphere with a diameter of almost half a mile (2,450 feet). So each gigantic particle, suspended in water, experiences a continuous barrage of irritating pricks at virtually every point over its surface as tiny water molecules hit and rebound from it.

Each time a water molecule strikes the suspended particle, the particle is given a push. To be sure, it is a tiny push because the particle is so massive compared to the water molecule. On average, these tiny pushes are incident at all points over the particle's huge surface with the result that the total push in any one direction is balanced by the total push in the opposite direction. With no net push, the particle sits still. But here is where fluctuations make a difference. With regularity, thought Einstein, fluctuations occur and the liquid molecules near the surface of the pollen particle move in an atypical fashion; that is, a group of molecules get bunched together in a tight little package and, like a convoy of trucks, they all move together in the same direction directly toward a small region on the surface of the suspended particle. When the packet of molecules collides with one small area on the particle's surface, the particle experiences an atypical push—both larger and more unbalanced in one particular direction. As a result, the particle moves in that direction.

A departure from the normal way water molecules move, a fluctuation in their motion, could move the suspended particle. That is what Einstein believed. Fluctuations themselves are random, so that the net pushes from the impacts of bunched liquid molecules are random—first in one direction, then in another, next in still another direction. The resulting motion of the suspended particle is a random, zig-zagging motion through the liquid.

It is not known exactly what mental images guided Einstein's thinking as he began his May paper on Brownian motion, of course, but this general picture is plausible. What is known is that Einstein wanted to develop a theory of Brownian motion that would expose the atomic nature of the liquid and that could be tested experimentally—quantitatively tested. If you imagine peering through a microscope to watch the zig-zagging motion of a suspended particle, what would you measure? Here again, Einstein's genius displays itself. And the process "was like a conjuring trick."[4] A few scientific papers, not many, seem like magic. Einstein's May paper is magic.

Looking ahead, Einstein must have thought carefully about how a theory that describes the motion of a particle suspended in a liquid could be confirmed. He did not design a theory that would have required the length of each little zig and each little zag to be measured. He may have recognized that such a measurement would have been exceedingly difficult. He did not design a theory that would have required the velocity of the pollen particles to be measured as they move in their sequential zigs and zags. Again, he may have recognized that that would be impossible. (Whether he knew it or not, attempts had been made to measure the velocity of Brownian particles. The attempts failed.) Einstein focused on something different.

Einstein begins by showing that size makes no difference. A liquid with suspended particles (like "large" particles of pollen) could, for his purposes, be treated just as a liquid with dissolved particles (like "small" sugar molecules). He did this by showing that the osmotic pressure was the same for molecules dissolved in a liquid (as in his dissertation) or particles suspended in a liquid (like pollen particles). "This consideration demonstrates that . . . according to this theory [the molecular-kinetic theory of heat], at great dilutions numerically equal quantities of dissolved molecules and suspended particles behave completely identically."[5] By showing that dissolved sugar molecules and suspended pollen particles could be treated similarly, he was able to import work he had undertaken for his dissertation, completed ten days earlier.

Next, Einstein considers the diffusion of the pollen-like particles suspended in the liquid in two ways. First, he employs an approach similar to the one he had used in his dissertation, that is, he considers diffusion as the movement of particles from a high concentration region to a lower concentration region. When the concentration of the suspended particles is uniform throughout the liquid, the diffusion ceases. (This is equivalent to going from a low-entropy condition to one of high entropy.) Opposing the diffusive motion of the pollen-like particles is the viscosity of the liquid. By taking these two opposing influences into account, Einstein determines the amount of pollen-like particles transported through the liquid. Specifically, he ends up with a diffusion coefficient identical to the one he had obtained in his dissertation.

Einstein's second approach is to consider diffusion as the consequence of the random motion of the suspended particles. However, as Einstein implies, the random motion of the pollen-like particles is due to the random motions of the surrounding liquid molecules. In any small increment of time, a pollen particle is pushed a tiny distance by a gang of water molecules. Of course, during the same time period, another pollen particle, located elsewhere in the liquid, is also pushed some tiny distance. In a random fashion, particles throughout the liquid are pushed this way and that. In the same random fashion, however, any one particle gets pushed again and again, which forces it along a zig-zagging path. Einstein develops a way to calculate the average distance moved by the pollen particles in an increment of time.

Einstein's final result was an expression for the average direct distance, from start to finish, that a particle would meander as it zig-zagged along over a particular period of time. Actually, Einstein simplified his final result, making it easier to test experimentally. He reduced his final result to one dimension, which meant that he expressed it in terms of the average distance a particle would travel *horizontally* in a particular time. Instead of a result that required the length of one zig or one zag to be measured, Einstein produced a result that required only the horizontal distance of many zigs and zags to be measured. The simplicity of this result endows it with the aura of magic.

That result was expressed in terms of the temperature of the liquid, T, the radii of the suspended particles, r, Avogadro's number, N, and a constant, k, now known as Boltzmann's constant, but recognized in 1905 as very significant. The last two constants, N and k, were particularly provocative because they had direct links to atoms.

At the end of his paper, Einstein used his result to show that in water (at 17° Celsius [C]) particles with a diameter of 0.001 mm would move a mean horizontal distance equal to 0.006 mm in one minute. That was what Einstein's theory of Brownian motion predicted—a very specific prediction. And that posed a challenge to physicists. Since Einstein knew and believed that theoretical predictions meant little if they were not tested experimentally, he closed his May paper with the exhortation, "Let us hope that a researcher will soon succeed in solving the problem posed here, which is of such importance in the theory of heat."[6]

Magic? With just an eye to a microscope, we could, in 1905, have observed the jigging and jogging of a suspended particle, which was direct evidence of molecular bombardment playing itself out in the surrounding liquid. With a stopwatch, we would have then been equipped to measure the horizontal distance traveled by the particle in a certain time period. By doing so, we would not only have tested Einstein's theoretical prediction but also have been able to announce with unprecedented confidence the verdict on atoms, the verdict on statistical fluctuations, and, depending on the last answer, the validity of classical thermodynamics.

The Response

One response to Einstein's May paper came in a conversation with Richard Lorenz. During this conversation, Einstein recounts, Lorenz made the suggestion that "many chemists would welcome an elementary theory of Brownian motion." "Responding to his request," Einstein tells us, "I present in the following a simple theory of this phenomenon."[7] Einstein finished his "simple theory" on April 1, 1908. In this 1908 paper, "Elementary Theory of Brownian Motion," Einstein describes his thinking in much more picturesque detail than he did three years earlier in his famous May 1905 paper. With phrases like "random motions of dissolved particles," "haphazard meandering of the molecules," and "this displacement [of the suspended particles] is determined only by the surrounding solvent," Einstein employs images much like those described earlier.

A more significant response, however, came from an experimentalist. Einstein's directive to experimentalists was this: take particles with a diameter of 0.001 mm, put them in water at 17° C, and observe, with a microscope, one of the particles for one minute. Record the horizontal distance it moved. Repeat the same procedure for several particles and average the recorded horizontal distance moved. The average horizontal distance, predicted Einstein, should be 0.006 mm. If the prediction was verified, it would provide compelling, almost incontrovertible evidence for atoms. The person who provided the confirmation most successfully was the French physicist Jean Baptiste Perrin.

Perrin was an atomist even before 1905, but more than that, by the time he became aware of Einstein's May paper, he was thoroughly familiar with Brownian motion. In 1908, Perrin carried out a series of experiments on Brownian motion whose purpose was to check the validity of Einstein's prediction. The outcome was immediately clear: "Right from the first measurements," wrote Perrin, "it became manifest, contrary to what might have been expected, that the displacements verified at least approximate the equation of Einstein."[8]

Perrin completed two series of experiments. The first, carried out with his doctoral student, Chaudesaigues, was conducted with particles of gamboge, a yellow vegetable latex. Perrin had learned how to prepare particles of gamboge with "exactly known diameter."[9] Working first with "relatively large granules of gamboge, of radius about 0.45 μ [0.00045 mm]" and later with "granules of radius equal 0.212 μ [0.000212 mm],"[10] Chaudesaigues recorded the position of the granule every thirty seconds for two minutes. Liquid solutions of both water and sugar water were used.

In a second series of experiments, this one in collaboration with Dabrowski, particles of mastic were substituted for gamboge. Taking turns at the microscope, Perrin and Dabrowski watched the mastic particles with radii of 0.0052 mm and, with the focus on one particular particle, its position was recorded every thirty seconds. The figure shows the paths of three different particles of mastic made by drawing a straight line between their thirty-second-interval positions.[11] When the experiments were complete and Perrin's data were analyzed, the results, wrote Perrin, "cannot leave any doubt of the rigorous exactitude of the formula proposed by Einstein."[12]

Einstein was delighted. He wrote to Perrin on November 11, 1909: "I wouldn't have thought it possible for the Brownian movement to be investigated with such precision; it is a piece of good luck for this subject that you undertook to study it."[13] Perrin was also delighted, then and later. In 1926, eighteen years after he finished his study of Brownian motion, he received the Nobel Prize.

Once Einstein's prediction was verified by Perrin, the atomic skeptics one by one capitulated. For example, around 1908, Henri Poincaré wrote that "the atomic hypothesis has recently acquired enough credence to cease being a mere hypothesis. Atoms are no longer just a useful fiction; we can rightfully claim to see them, since we can actually count them."[14] Perhaps the captain of all the skeptics was Wilhelm Ostwald, but his battle

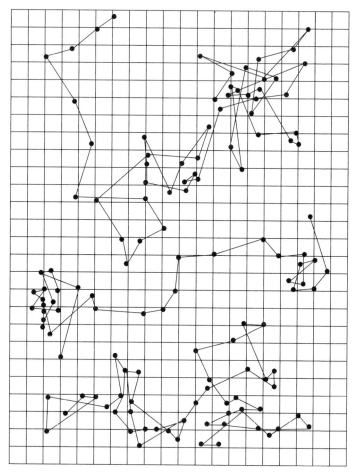

Brownian motion. The positions of three granules of mastic were recorded every thirty seconds by Jean Perrin. Each thirty-second position was connected by a straight line, which reveals the zig-zagging path of the granule.

against atoms was lost. In 1909, Ostwald wrote, "I have convinced myself that we have recently come into possession of experimental proof of the discrete or grainy nature of matter, for which the atomic hypothesis had vainly sought for centuries, even millennia."[15]

How about classical thermodynamics? In the beginning sentences of the May paper, we are presented with two alternatives: if Brownian motion is observed, as predicted, then "classical thermodynamics can no longer be viewed as strictly valid." A positive way of saying the same thing would be: if Brownian motion is observed, Boltzmann's probabilistic interpretation of entropy must be valid. In 1911, at the First Solvay Conference in Brussels, Einstein mentioned Boltzmann's Principle, which is a statistical interpretation of entropy, and said, "We should admit its validity without any reservations."[16] Six years later, in 1917, Einstein again made his position clear: "Because of the understanding of the essence of Brownian motion, all doubts vanished about the correctness of Boltzmann's interpretation of the thermodynamic laws."[17] Boltzmann interpreted the thermodynamic laws to be true only in the statistical sense.

Max Born put it all together in 1949: Einstein's theory of Brownian motion did "more than any other work to convince physicists of the reality of atoms and molecules, of the kinetic theory of heat, and of the fundamental part of probability in the natural laws."[18] Most were persuaded, but not all. In 1909, Ernst Mach, whose work had been an early influence on Einstein, wrote an essay in which he once again separated his views from those who believed in atoms. Mach sent his essay to Einstein. Einstein responded on August 9, 1909, apparently in a way he thought would influence Mach. In his letter, Einstein wrote: "Since I cannot think of any other way in which to show you my gratitude, I am sending you some of my papers. I would especially like to ask you to take a cursory look at the one on

Brownian motion, because here is a motion that we believe must be interpreted as 'thermal motion.'"[19]

It is noteworthy that Einstein singled out his paper on Brownian motion in the hope that it might be the basis for Mach's acceptance of the atomic theory. The likelihood is that in this, Einstein failed. Mach died in 1916, eleven years after the May paper. Mach may have been the last scientist to deny the existence of atoms. Whether he ever really came to believe in their reality is uncertain, but Einstein's May paper did not do for Mach what Einstein had hoped; namely, it did not immediately convert Mach into an atomist. For the rest, however, Einstein's Brownian motion provided all the necessary evidence. Soon after May 1905, questions about the reality of atoms were no longer heard.

On the Electrodynamics
of Moving Bodies

A. Einstein

The June paper was received by the editor of *Annalen der Physik* on June 30, 1905, and published that same year in *Annalen der Physik*, volume 17, pages 891–921.

Albert Einstein in 1921 or 1922 at the height of his world fame. When a prediction of Einstein's general theory of relativity was confirmed in 1919, Einstein became a celebrity.

The Merger of Space and Time

In his "Autobiographical Notes," Einstein identified two criteria that a physical theory must meet. The first is obvious: "the theory must not contradict empirical facts."[1] The second criterion is less obvious: it concerns "the 'naturalness' or 'logical simplicity' of the premises" of the theory. The first criterion, continues Einstein, "refers to the 'external confirmation'" of the theory and the second is concerned "with the 'inner perfection' of the theory."[2]

Agreement between theory and experimental facts is regarded as absolutely essential to all science. Sometimes the facts precede theoretical explanation. This was the case, for example, with Brownian motion. Particles, suspended in a liquid, had been observed zig-zagging through the liquid in a vigorous fashion for decades before Einstein's May paper provided what came to be a quantitative explanation. At other times, the theory precedes the facts, which was the case with Einstein's March paper, in which his theory identified some physical properties of the photoelectric effect before experimentalists had established them as facts. In this latter example, as is frequently the case, theory identified what facts can be observed; sometimes theory actually *defines* the facts that can be observed. Werner Heisenberg, who approached his creation of quantum mechanics in terms of what could actually be observed, was surprised when Einstein asserted, "It is always the theory which decides what can be observed."[3] Whatever the sequence, there must be harmony between theory and experiment. If there is disagreement and if that disagreement persists, experiment becomes, as always, the final arbiter and the theory must be brought into agreement with experiment.

It is less clear what Einstein's second criterion, "inner perfection," means. Clearly, a one-fact theory, specifically created to explain one particular fact, is not a good theory. A theory whose premises are arbitrary and convoluted lacks perfection. Einstein acknowledged that a precise definition of "inner perfection" might not be possible. Yet he absolutely and precisely knew what he meant by "inner perfection." To say that something has "inner perfection" is expressing a judgment that is both objective and subjective. Yet such judgments can be and are made routinely. Consider, for example, Paul Dirac's description of Einstein's general theory of relativity which, he said, "has a character of excellence of its own. Anyone who appreciates the fundamental harmony connecting the way Nature runs and general mathematical principles must feel that a theory with the beauty and elegance of Einstein's theory *has* to be substantially correct."[4] Dirac knew what "inner perfection" meant. In fact, a great physical theory exudes a perfection for those prepared to see it.

Einstein's June paper on the special theory of relativity was at the high end of perfection. It radiates perfection.

The Setting

In 1905, three great physical theories were used to account for the variety of ways that Nature conducts itself. The first of these great theories, called mechanics, deals with motion and energy. The second, thermodynamics, also deals with energy, but more specifically thermal energy and other thermal phenomena—heat and temperature. During the second half of the nineteenth century, however, thermodynamics had, in principle, been subsumed by Newtonian mechanics, as thermal phenomena could be explained in terms of the motions of unseen atoms. All things related to electricity and magnetism came into the domain of the third great theory, electromagnetism. These three great theories of physics had become highly refined and, in the waning years of the nineteenth century, were thought to be in, or near, their final forms. Nonetheless, as the twentieth century dawned, physicists around the world were an anxious lot.

The decade immediately preceding 1905 was a lively time in the physics profession. Recent, totally unexpected discoveries, including X-rays (1895), radioactivity (1896), the electron (1897), and the quantum (1900), had ripped the covers off the book of physics and had given physicists notice that Newton's mechanics and Maxwell's electromagnetism were not the final chapters. The Epilogue of physics, already written by some physicists, was consigned to the dustbin of history.

The new discoveries were just a part of the turmoil; a failed experiment added to the angst of physicists. In 1887, an experiment conducted in Cleveland, Ohio, failed to detect "the ether," and that failure challenged, if not negated, a commonly held assumption for which there were no alternatives. Desperate, even frantic attempts were made to transform the decisive failure into an acceptable success. In the context of 1905, Einstein's June paper had an element of timeliness. Some of the ideas that came out of Einstein's June paper were "in the air" in 1905 and had been floating around for several years prior. Though his paper was timely and used ideas in common currency, the form and outcomes of Einstein's June paper strained the credulity of physicists.

Physicists thought they knew that light propagated from the Sun to Earth as a wave. But for this to happen, there had to be a medium that, effectively, propelled the light wave. Sound is a wave and as a wave it requires a medium. Virtually all material objects can serve as a medium to transmit sound waves; however, it is air that propagates sound waves from a mother's mouth to a child's ear. Sound waves cannot propagate in a vacuum. Just as with sound, and for all other wave phenomena, it was believed with absolute confidence that light waves required a medium. And not just any medium would do: light required a most unusual medium. That medium was the ether (sometimes spelled aether).

What made the medium for light waves unusual? To begin, light arrives at Earth from far-off stars and galaxies, which themselves are, to the best of our knowledge, distributed throughout the entire universe; therefore, the medium that propelled light had to fill the vastness of all space. Further, since light propagates from place to place at the unearthly speed of 186,000 miles per second (mi/s), the medium had to have the rarest of properties. Finally, Earth and the other planets plow through this pervasive medium as they orbit around the Sun and they do so without any observable effects; for example, after millions of years pushing its way through the medium, Earth's surface has not been "sanded" smooth nor has Earth's orbital speed slowed because of any retarding influences caused by the medium. Whatever its properties, the omnipresent medium, the ether, did not affect either the planets or their motions. The ether was indeed unusual.

The ether, as the medium for light, gave meaning to the speed of light. When the speed of an aircraft is given as 520 miles per hour (mi/h), it goes without saying that the speed is given relative to Earth, that is, relative to an Earth-based coordinate system. Speed has meaning only when it is referred to a particular coordinate system. The coordinate system with respect to which light had its 186,000 mi/s speed was a coordinate system attached to the ether.

A coordinate system, also called a reference frame, is also required to specify the locations of objects or events. If, for exam-

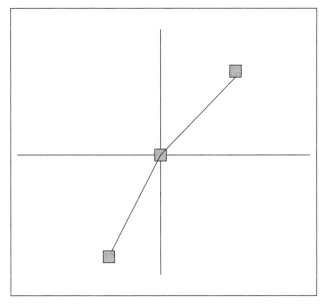

A coordinate system centered on New York City. In this coordinate system, Washington, D.C., is located 220 miles SSW. In this New York–centered coordinate system, Boston is located about 220 miles NNE of New York.

ple, a New Yorker says that Washington, D.C., is 220 miles south southwest, the coordinate system being used is attached to Earth; it has its origin fixed in Manhattan, and has one axis pointing north-south, another east-west, and the third axis up-down. In terms of that coordinate system, New Yorkers can locate any city on Earth. A location has no meaning in the absolute sense. A New Yorker may say that Washington, D.C., is 220 miles south southwest, and a New Orleanian may say that Washington, D.C., is 1,080 miles north northeast. Both statements are true. Location is always relative to something else.

A special kind of coordinate system, called an inertial coordinate system, played the lead role in Einstein's June paper. An inertial coordinate system is defined by Newton's First Law of Motion, which equates rest and uniform motion (motion with constant velocity, which is motion in a straight line at a constant speed). How can rest be equated with uniform motion? Consider, for example, a cup of coffee sitting at rest on the front porch of a farmhouse in western Kansas. The front porch of the farmhouse is an inertial coordinate system which is at rest with respect to Earth. A second cup of coffee sits on a passenger's serving tray in a Boeing 767, flying due east in smooth air at a level altitude of 37,000 feet at a constant speed of 580 mi/h. The Boeing 767 is an inertial coordinate system moving at a constant velocity with respect to Earth. Newton's First Law of Motion says that there is no way to look at the two cups of coffee, each in their own respective inertial coordinate system, and decide which is at rest or which is moving. In fact, no experiment can identify which coffee cup is moving. Rest and uniform motion are indistinguishable. (Of course, if the 767's uniform motion is interrupted by turbulent air, the sloshing coffee makes it clear which cup is moving.) Because there is no experimental way to distinguish between inertial coordinate systems at rest and in uniform motion, a principle came out of Newtonian mechanics called the relativity principle, which simply expressed the fact that all inertial coordinate systems are equivalent. Inertial coordinate systems, or inertial reference frames, were important in Galilean and Newtonian physics; however, as indicated, they occupy center stage in Einstein's special theory of relativity.

Since the ether filled the entire universe, it could not move from one place to another. In other words, the ether was at absolute rest. (Of course, absolute rest has no meaning except when it relates to a particular coordinate system. In this example, the ether was deemed to be at rest with respect to the universe as a whole.) The ether was an inertial coordinate system at rest in the universe and in that system, light traveled at 186,000 mi/s.

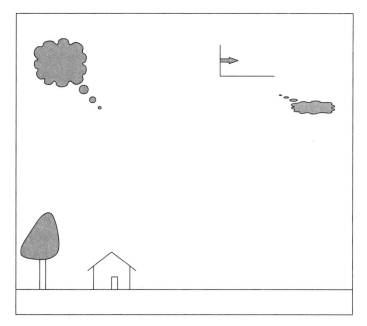

The Kansas farmhouse has a coordinate system fixed to the Earth. In this co-ordinate system, the plane is 37,000 feet above Earth and is moving east at 580 mph. The plane's coordinate system is fixed to the plane. For passengers looking out the plane's windows, the surface of Earth appears to move 580 mph to the west. In the plane's coordinate system, passengers and their coffee cups are at rest.

Light travels at a fixed speed through the ether, the ether surrounds Earth, and Earth moves in its orbital motion around the Sun passing through the ether, which is at absolute rest. These facts, believed by Einstein's 1905 contemporaries, provided the basis for an experiment physicists designed to compare the speed of light as it moves through the ether in two opposing directions. Here is such an experiment: First, we measure the speed of light moving through the ether in the same direction as Earth's orbital motion and then we repeat the measurement as the light moves in the direction opposite to Earth's orbital motion. In the first measurement, Earth is moving alongside the light as it propagates through the ether and in the second, it is moving into the oncoming light. We would expect the second measured speed of light to be greater than the first. By measuring the difference between the two speeds, Earth's speed through the ether can be determined.

This was the kind of experiment that Albert A. Michelson and Edward W. Morley completed in 1887. When they compared the speeds of two light beams sent through the ether in different directions relative to the direction of Earth's orbital motion, they expected them to be different. They were not. There was no difference between the two speeds. Physicists who thought about it were certainly surprised.

The experiment was deemed a failure—one of the most famous failed experiments in the history of physics. Even though the experiment was repeated at different times, at different locations, and at different altitudes, the result was the same: no difference in speed.

Physicists, to their credit, accepted the result, albeit grudgingly. The failed experiment, which negated the accepted dogma of a static ether, implicitly raised questions about the ether concept itself. But the ether was needed to propagate light. The great physicist Hendrik A. Lorentz wrote to Lord Rayleigh on August 18, 1892: "I am utterly at a loss to clear away this contradiction [between the ether theory and the result of the Michelson-Morley experiment], and yet I believe if we were to abandon Fresnel theory [the idea that the ether was at rest, but that some ether was dragged within objects moving through it], we should have no adequate theory at all . . . Can there be some point in the theory of Mr. Michelson's experiment which has as yet been overlooked?"[5] Physicists, acting out of a sense of desperation, responded to the crisis.

In short, the experimental result had to be accepted, but the ether had to be retained. Therefore, a series of ad hoc remedies were proposed to make the result of the Michelson-Morley experiment compatible with a static ether. Perhaps the most bizarre suggestion was made by Lorentz in 1892, and independently by George F. FitzGerald in 1889. They proposed that one dimension of a moving object, by virtue of its motion through the ether, would contract; specifically, that the one dimension parallel to the object's line of motion through the ether would contract. According to this proposal, a steel bar aligned along the direction of Earth's motion, and moving with Earth through the ether, would contract. In 1899 and 1904, Lorentz adopted some assumptions, and on the basis of these assumptions, he developed a set of equations (later called the Lorentz transformation equations) that linked observations between stationary and moving inertial coordinate systems. In the process, he showed that the Lorentz-FitzGerald length contraction was consistent with these transformation equations.

These strange ideas were invented to patch over the problems engendered by the ether concept. The patches were offensive, but physicists believed the ether was required in order for light to travel from place to place. In addition to ad hoc remedies, basic ideas were looked at afresh. In 1898, for example, Henri Poincaré raised questions about time: "We have no direct intuition about the equality of two time intervals. People who believe they have this intuition are the dupes of an illusion."[6] And in 1904, at the St. Louis World's Fair, Poincaré asked, "What is the aether, how are its molecules arrayed, do they attract or repel each other?"[7] During his remarks about time, Poincaré talked about clock synchronization; Lorentz defined "local time," which Poincaré elaborated further; Poincaré brought the Galilean-Newtonian relativity principle into the discussions.

Ironically, the weird idea of length contraction as well as new insights into time and clock synchronization were pivotal parts of Einstein's June paper, but with a profound difference. Instead of coming out of the blue, length contraction and time considerations emerged as logical consequences of the two principles that were the basis of the special theory of relativity.

Provocative thoughts related to his June paper had been in Einstein's mind for several years. In August 1899, for example, Einstein wrote a letter to his future wife Mileva in which he expressed his doubts as to whether the ether had any physical meaning. At another moment he imagined himself running alongside a light wave and, in this image, he identified several problems. Over many months, these thoughts and others faded in and out of Einstein's conscious thoughts; they became clearer, and, to some extent, they were arranged in his mind. These musings were pieces of what would eventually become a coherent whole. All that was needed was one flash of insight. That insight came in the spring of 1905 during a visit to his friend, Michele Besso. Here is Einstein's account, given in 1922, of that dramatic moment:

> Unexpectedly a friend of mine in Bern then helped me. That was a very beautiful day when I visited him and began to talk as follows: "I have recently had a question which was difficult for me to understand." Trying a lot of discussions with him, I could suddenly comprehend the matter. Next day I visited him and said to him without greeting: "Thank you. I've completely solved my problem." My solution was really for the very concept of time, that is, that time is not absolutely defined but there is an inseparable connection between time and the signal velocity. With this conception, the foregoing extraordinary difficulty could be thoroughly solved. Five weeks after my recognition of this, the present theory of special relativity was completed.[8]

Two years later, in 1924, Einstein described this same moment of insight in more revealing words:

After seven years of reflection in vain [1898–1905], the solution came to me suddenly with the thought that our concepts and laws of space and time can only claim validity insofar as they stand in a clear relation to our experiences; and that experience could very well lead to the alteration of these concepts and laws. By a revision of the concept of simultaneity into a more malleable form, I thus arrived at the special theory of relativity.[9]

Einstein's earlier description, "an inseparable connection between time and the signal velocity," and his later words, "by a revision of the concept of simultaneity," make clear his insight. It was that the commonly held understanding of simultaneity was fundamentally flawed. Implicitly, it is assumed that if one person observes two events to be simultaneous, all other observers will agree that they are simultaneous. Einstein saw the fallacy in this and once this insight came, all the pieces fell together into a logical whole. Five weeks later, Einstein completed his special theory of relativity.

The June 1905 Paper

There is a disturbing fact about current science. If a contemporary journal editor received a manuscript from an unknown clerk working in a nondescript place, that editor would, in all likelihood, reject it, possibly without even reading it. Yet Einstein, an unknown clerk working in a nondescript place, produced one of the most significant and beautiful manuscripts in the history of physics. A prominent Einstein scholar, Arthur Miller, writes about this paper:

Page for page, Einstein's relativity paper is unparalleled in the history of science in its depth, breadth, and sheer intellectual virtuosity. Einstein developed one of the most far-reaching theories in physics in a literary and scientific style that was parsimonious yet not lacking in essentials; in a pace that, where necessary, possessed a properly slow cadence yet was presented in thirty pages of print, is developed almost like an essay. Written in white heat in about five weeks, it is pristine in form, and yet in its own way as complete as Newton's book-length *Principia*.[10]

Einstein's June paper begins with this sentence: "It is well known that Maxwell's electrodynamics—as usually understood at present—when applied to moving bodies, leads to asymmetries that do not seem to be inherent in the phenomenon."[11] Once again, as in the March paper, Einstein opens with an apparent contradiction. There are different ways to illustrate the asymmetries that Einstein refers to in his opening sentence. All of the ways, however, involve two inertial coordinate systems. In one system there is some configuration of objects that can include a charged object, a conductor, and a magnet, along with an attending observer. The second coordinate system with its observer is in motion relative to the first system.

To illustrate the asymmetries, Einstein employed currents resulting from the relative motion between electrical conductors and magnets. There are other ways, and simpler ways, to demonstrate Einstein's basic concern. Consider, for example, an inertial coordinate system at rest. In this rest coordinate system are an observer, Observer *A,* and a charged object. With a sensitive compass the observer examines the space around the charged object and finds no evidence of a magnetic field. Next consider an inertial coordinate system with another observer, Observer *B,* moving with uniform motion to the east. Observer *B* sees the charged object moving to the west and, with a sensitive compass, detects a magnetic field in the space around the moving charged object. From the perspective of one inertial coordinate system there is no magnetic force acting on the compass pointer; from the perspective of another inertial coordinate system there *is* a magnetic force that aligns the compass needle. If, as assumed, all inertial coordinate systems are equivalent, there is an obvious contradiction, or as Einstein stated, an asymmetry: one cannot have forces producing physical changes in one inertial system and no forces and no physical changes in another. In short, either there is or there is not a magnetic field. These electromagnetic experiments were at odds with the relativity principle.

After Einstein establishes an example of an asymmetry (the first paragraph), he moves quickly. He asserts that, in terms of physics, the idea of absolute rest has no meaning. By this assertion he rids physics of the ether. Einstein renders the ether, at absolute rest, "superfluous." He next asserts that the laws of electromagnetism, optics, and mechanics are valid in all inertial reference systems and he raises the principle of relativity to the status of an axiom. He immediately introduces a second principle, only "seemingly incompatible," that also carries the status of an axiom, namely, that the speed of light is the same in all inertial reference systems. In concise form, the two principles of special relativity that appear in the second paragraph of his June paper are:

The Principle of Relativity—The laws of physics are the same in all inertial reference systems or, by means of physical experiments, one inertial coordinate system cannot be distinguished from another inertial coordinate system.

The Principle of the Constancy of the Speed of Light—The speed of light is the same in all inertial reference systems, independent of the speeds of either the source of the light or the detector of the light.

By the end of the second paragraph, Einstein has put all his cards face up on the table. In the following twenty-eight pages he systematically develops the consequences of the two principles for mechanics (Part I) and for electromagnetism (Part II).

Despite the two principles' rather ho-hum appearance, in the context of 1905, they were mutually inconsistent. It was only in the coordinate system of the ether that light had a fixed speed; in other inertial coordinate systems, the observed speed of light would depend, for example, on whether the observer is moving toward the oncoming light or away from the oncoming light. Einstein warned the reader of his June paper that the two postulates were only "seemingly incompatible." The apparent innocence of the two principles completely disappears when they are brought together and the logical consequences begin to emerge. With the consequences, the two principles become the basis for a startling new world.

Einstein prepares the way by sharpening the concept of time. The time ascribed to a particular event is determined by a reading on a clock.

> We have to bear in mind that all our propositions involving time are always propositions about *simultaneous events*. If, for example, I say that "the train arrives here at 7 o'clock," that means, more or less, "the pointing of the small hand of my clock to 7 and the arrival of the train are simultaneous events."[12]

However, Einstein points out that in this case, the clock and the arrival of the train are essentially at the same location. When times have to be ascribed to events at locations that are separate from one another, careful attention must be given to how that is done. Imagine an inertial coordinate system at rest and further imagine that one clock is used to record the time of both nearby and far-off events. The immediate problem is that the times recorded by the one clock depend on where the observer and her clock are located in the coordinate system. Einstein's solution was to imagine a clock at the location of each event, say locations A and B. The time of an event at A is determined by the clock at A and the time of an event at B is determined by the clock at B. Then, by following a well-described procedure to synchronize the clocks at A and B, the observer at location A can know the time of an event at location B. After this brief discussion, Einstein asserts,

> The "time" of an event is the reading obtained simultaneously with the event from a clock at rest that is located at the place of the event and that for all time determinations is in synchrony with a specified clock at rest.[13]

By sharpening definitions, Einstein is laying the groundwork piece by piece.

Next, based on the two principles of special relativity plus his prescription for determining the time of an event, Einstein gives what is essentially a qualitative discussion designed to show how two different observers would measure the length of an object. Again we are called upon to use our imaginations.

Imagine the first observer, Observer A, and a steel beam in an inertial coordinate system that moves east uniformly with respect to Earth. The beam is oriented due east and west, along the direction of motion. Observer A, at rest *with respect to the steel beam*, would measure the length of the beam by laying out a ruler along the beam's length. Call this measured length the rest length, L_{rest}, because it is measured by an observer at rest with respect to the steel beam.

Now imagine a second observer, Observer B, in a second inertial coordinate system that is at rest with respect to Earth. This stationary observer sees the steel beam moving east. The stationary observer with stationary synchronized clocks measures the length of the beam by determining the exact locations of the two ends of the beam at the same instant of time. For example, if we want to measure the length of a car as it moves down the road, we would mark the location of the leading edge of the front bumper at the same instant as we locate the trailing edge of the

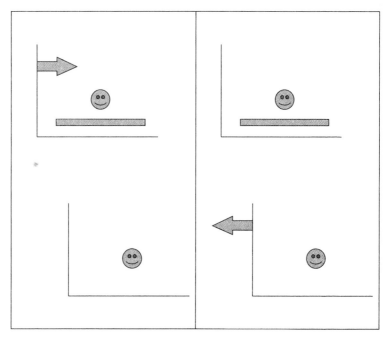

Observer B (lower left) sees the steel beam and Observer A (upper left) moving east. The beam is at rest relative to A and A measures the length of the beam to be L_{rest}; the beam is moving relative to B and B measures the length as L_{moving}. L_{moving} is less than L_{rest}. Observer A (upper right) sees Observer B moving west.

back bumper and measure the distance between these two locations. Since for this observer the beam is moving, we call this measured length the moving length, L_{moving}.

Common sense says that the two lengths, L_{rest} and L_{moving}, are the same. We will find, wrote Einstein, that the two lengths are *not* equal. This may sound contradictory; however, there is more to the story. Time again enters. The Earth-based observer, Observer B, believes that he located the two ends of the eastward-moving steel beam at the same instant, that is, simultaneously. The observer moving with the beam, Observer *A*, might say, "Excuse me, you located the two ends of the beam at different times, *not* simultaneously." Einstein concludes,

> Thus we see that we must not ascribe *absolute* meaning to the concept of simultaneity; instead, two events that are simultaneous when observed from some particular coordinate system can no longer be considered simultaneous when observed from a system that is moving relative to that system.[14]

At this point, Einstein had taken steps to prepare the reader to rethink comfortable, common-sense ideas; namely, that the length of a rigid object is *not* the same for all observers and that the simultaneity concept also depends on the observer. He could have said more: he could have added that Newton's concept of absolute time was a mistaken idea.

From this largely qualitative discussion, Einstein brings in some mathematics. Once again imagine two inertial coordinate systems, each with an observer—one at rest with respect to Earth, a second moving uniformly to the east. Assume an event: a bolt of lightning strikes a fencepost. The observer in the rest system records the specific location and the particular time of the event. The observer in the moving system also records the exact location and precise time of the event. Two sets of location and time measurements exist: one relative to the rest system and the other relative to the moving system. What mathematical equations connect these two sets of recorded data? Or, to ask the question another way, how do you transform the measurement data of one observer into the measurement data of the other observer? Einstein used his two principles, the relativity principle and the constancy of light principle, to develop equations that responded to these questions and ended with the same equations as those that Lorentz derived in 1895 and 1899. These are the same equations, but with a difference: Lorentz's equations resulted from his ad hoc attempt to explain the "failed" Michelson-Morley experiment; Einstein's were a direct consequence of his two principles and his conclusions about time.

Einstein uses these transformation equations to show that whether or not an observer is stationary with respect to a beam of light or moving alongside a beam of light, the speed of light is the same for both. "This proves," writes Einstein, "that our two fundamental principles are compatible."[15]

With these transformation equations in hand (today called the Lorentz transformation equations), Einstein returns to his analysis of length and time. Length, he indicates, is relative. One dimension of an object, the one that is parallel to the direction of motion, contracts. Other dimensions are unaffected. A steel beam that is measured to be ten feet long by an observer standing beside the beam (at rest with respect to the beam) is measured to be less than ten feet by an observer who sees the beam moving by him. For example, if a 120-inch (ten-foot) steel beam is moving at a speed equal to 10 percent the speed of light, or 18,600 mi/s, it contracts by 0.6 inches; if it is moving at a speed equal to 25 percent the speed of light, or 46,500 mi/s, it contracts by 3.8 inches. At half the speed of light, or 93,000 mi/s, it contracts by 16.1 inches, about 13 percent of its length. Clearly, the size of the contraction becomes significant only at speeds approaching the speed of light. At the speed of light, the contraction is 100 percent. The length of the beam contracts to zero. At speeds greater than the speed of light, the beam's length would become negative. Since a negative length makes no sense at all, Einstein asserts that the speed of light is Nature's speed limit. Nothing can move faster than light.

Time is also relative. Imagine two inertial coordinate systems, each with clocks. Both clocks keep perfect time and agree with each other when both coordinate systems are at rest. If we put one inertial coordinate system, along with its clock, in motion, when the moving clock is viewed from the rest system, it lags behind the clock at rest. This is not some weird malfunction of a clock. This has to do with time itself. Clocks measure time and moving clocks lose time relative to clocks at rest. This property of time means that if, in a rest system, there are two synchronized clocks at locations A and B, and if the clock at B is transported, at a speed v, to the clock at A, the two clocks will no longer be synchronized—clock B will be behind clock A. Or if both synchronized clocks are located together at A and if one clock is taken around the world, upon arrival back at A, the traveling clock will have lost time relative to the stationary clock. (This has been verified by flying an atomic clock around the world and, upon return, comparing it to the stay-at-home clock.)

Clocks come in many shapes and forms. A person is a clock. An individual's metabolic rate, just like a wrist watch, keeps time and announces the approach of lunch time. Some kind of internal clock governs the aging process. All this to say that if a mother were to leave home on a very long, unimaginably high-speed trip, her internal clocks would slow relative to those left behind and the mother could return home younger than her stay-at-home son. Einstein's relativity theory revealed strange things about Nature.

Einstein ends Part I of the June paper with a discussion of velocities. Always working from the two principles, Einstein shows that no two velocities can combine so as to exceed the speed of light. Once again we must use our imagination. Imagine an observer standing by a long, straight railroad track built on a vast level plain. Next imagine a train, with one very long flatcar, moving at a speed of 111,600 mi/s (60 percent the speed of light) due east. Next, imagine a car driving east along the top of the flatcar at a speed, relative to the flatcar, of 111,600 mi/s (again, 60 percent the speed of light). Common sense says that the observer, standing on the ground and watching the train and car pass, would see the car moving east at the sum of these two speeds, that is, 111,600 mi/s + 111,600 mi/s or 223,200 mi/s, which is 120 percent the speed of light. Not so if the logic of the two principles is accepted. Working from the two principles, Einstein showed that the observer along the tracks would see the car moving at a speed of 163,680 mi/s or 88 percent the speed of light. To repeat, the special theory of relativity sets a limit on how fast an object can go. The speed limit is 186,000 mi/s.

The results of Part I are intellectually and emotionally stunning. Absolute space and absolute time, the foundations of Newtonian physics, are seen to be a figment of our imagination. Absolute simultaneity is also a myth. Events seen as simultaneous by one observer are seen as events with a time interval between them by another observer. The time interval between two events occurring at different spatial locations is not an absolute: one observer might witness two events as simultaneous, another observer may say one event came five seconds before the other, a third observer may say that one event came ten seconds before the other. The time interval between two events depends on the inertial coordinate system of the observer. These results violate common sense and trouble emotions. Although the time interval between two events depends on the coordinate system, the relationship between cause and effect does not. Within the special theory of relativity causality is preserved, that is, cause *always* precedes effect. The time interval between cause and effect can vary with the inertial coordinate system, but the order cannot be reversed.

In Einstein's May 1905 letter to Conrad Habicht, the same letter in which he called his March paper "very revolutionary," Einstein refers to his imminent June paper by saying that it concerns the electrodynamics of moving bodies and is based on "a modification of the theory of space and time; the purely kinematic part of this paper will surely interest you."[16] Indeed, after June 1905, our understanding of our common-sense three-dimensional world was transformed to include Nature's strange four-dimensional universe in which the dimension of time is somehow merged with the three dimensions of space.

"[T]he purely kinematic part of this paper" is what Einstein presents in Part I of the June paper, the part that jars human sensibilities. The second part of the June paper applies the two principles to electromagnetism. Remember, the title of the paper is "On the Electrodynamics of Moving Bodies." It was the "asymmetries" in the domain of electromagnetism that existed between different inertial reference systems that, according to Einstein, "compelled" him to develop the special theory of relativity. In the second part, Einstein showed that his transformation equations (the Lorentz transformation equations) could be applied to the equations of electromagnetism and that they retained their validity in all inertial coordinate systems. He also showed that "the asymmetry mentioned in the Introduction when considering the currents produced by the relative motion of a magnet and a conductor, disappears."[17] Einstein's two principles, working in tandem, resolved the "asymmetries."

In this famous June paper, Einstein included no citations. Much of his source material was "in the air" among scientists in 1905, and some of these ideas had been published. Einstein could have cited the work of Lorentz and Poincaré; however, to do so would have been a bit artificial and perhaps even disingenuous. In the development of his special theory of relativity, Einstein did not draw from or build upon the work of others. He adopted two principles as axiomatic, and by means of his intellectual prowess, he brought the unseen consequences of the two principles into full view. At the end of the paper, he thanked his friend, Michele Besso.

The Response

Einstein, according to his sister Maja, expected quick reactions to his June paper:

> "The young scholar," wrote Maja, "imagined that his publication in the renowned and much-read journal [*Annalen der Physik*] would draw immediate attention. He expected sharp opposition and the severest criticism. But he was very disappointed. His publication was followed by icy silence. The next few issues of the journal did not mention his paper at all. The professional circles took an attitude of wait and see."[18]

The silence was brief, and when it was broken, Einstein might well have wished it to continue. There appeared to be problems. Walter Kaufmann, a very prominent experimental physicist, was the author of the first paper in *Annalen der Physik*, in 1906, that specifically mentions Einstein's June paper.

Just before Einstein expresses his thanks to Besso in his June paper, he concludes, "These three relations are a complete expression of the laws by which the electron must move according to the theory presented here."[19] Concerning the motion of electrons, these words left Einstein no wiggle room. Whether Einstein knew it or not, Kaufmann had been conducting experiments on the motion of fast electrons in electromagnetic fields since 1887 and he had experimental data. How did Kaufmann's data square with Einstein's theory? Kaufmann begins his 1906 paper with these words: "I anticipate right here the general result of the measurements to be described in the following: *the measurement results are not compatible with the Lorentz-Einsteinian fundamental assumptions*."[20] Kaufmann's stature was such that his experimental data attracted the attention of other physicists; in fact, two physicists had developed a physical theory to explain Kaufmann's experimental results, the first in 1902 and the second in 1904. In 1906, Einstein faced conflicting data and two competing theories.

Did Einstein rise to his own defense? He said little publicly until one year later. Ever confident in his own physics, Einstein, in 1907, responded by calling upon time and "a more diverse body of observations" before overruling his theory in favor of Kaufmann's data.[21] Einstein did, however, comment specifically on the two theories created to explain Kaufmann's data. His remarks reveal what, for Einstein, constitutes "inner perfection" of a physical theory: "In my opinion both [their] theories have a rather small probability, because their fundamental assumptions concerning the mass of moving electrons are not explainable in terms of theoretical systems which embrace a greater complex of phenomena."[22] It took until 1916 to identify the flaws in Kaufmann's experimental procedures; when corrected, the disparity between Kaufmann's data and Einstein's theory disappeared. Long before 1916, however, Einstein received a reassuring letter from Alfred Bucherer, who, like Kaufmann, was also conducting experiments on fast-moving electrons. In 1908 Bucherer wrote: "First of all, I would like to take the liberty of informing you that I have proved the validity of the relativity principle beyond any doubts by means of careful experiments."[23]

By 1906, in addition to Kaufmann's disturbing paper, Einstein was receiving a number of letters about his relativity theory, and physicists were coming to the Bern patent office to speak with him. By 1911, manuscripts on relativity submitted to *Annalen der Physik* had become so numerous that Max Planck, the advisor on theoretical physics to the journal's editor, suggested they be diverted to another journal.[24]

The response to Einstein's theory of relativity was intense. It was also messy. The particle theory of light, which Einstein proposed in his revolutionary March paper, was rejected outright by physicists, but they did so quietly. Little was said about Einstein's light quanta. They were more vociferous, however, in regard to the June 1905 paper because it disrupted the established order. Of all Einstein's 1905 papers, it is the June paper that stands apart in terms of the controversy it generated. Further, the responses occupy extremes. On the one hand, there were those who accepted the tight logic of the theory and regarded Einstein with high esteem. On the other hand, there were people who vigorously rejected the theory and regarded its creator with contempt, especially after 1919.

The reactions to Einstein's 1905 special theory of relativity were enhanced by his 1915 general theory of relativity which, after lying dormant during the Great War, regained prominence in 1919 when Arthur Eddington verified that a star's light was deflected when it grazed the Sun on its way to Earth—a key prediction of the general theory. Eddington's affirmation of the general theory launched Einstein into the public arena and, soon after, Einstein became the target of personal attacks.

From the outset, Einstein's special theory of relativity has stimulated the thoughts and fired the emotions of both supporters and opponents inside and outside the physics profession. As the editor of the *American Journal of Physics* from 1978 to 1988, I received scores of manuscripts from authors who attacked relativity theory and purported to reveal various errors that Einstein had made. Many of these manuscripts were driven as much by emotion as by intellectual considerations. Many find the consequences of the special theory too abstract and too contrary to common sense. Others accepted the theory but rejected its implications. Still others could not accept a theory that was, in essence, the product of pure thought rather than of hard experimental facts. Some simply could not give up the ether. Reactions varied from one national setting and intellectual climate to another.[25]

In 1911, in an official address, the president of the American

Physical Society expressed his discomfort with the consequences of the theory: "[B]y no stretch of my imagination can I make myself believe in the reality of the fourth dimension." He continued, "Can we venture to believe that the new space and time introduced by the principle of relativity are either thus intelligible now or will become so hereafter?"[26] In August 1920, a meeting was held in the largest concert hall in Berlin for the express purpose of attacking the theory of relativity and its author.[27] And in a *New York Times* editorial published on January 28, 1928, we read: "Tennyson claimed for faith the function of believing what we cannot prove. The new physics comes perilously close to providing what most of us cannot believe; at least until we have rid ourselves completely of established notions and forms of thought. Relativity translates time in terms of space and space in terms of time."[28]

Among most physicists, particularly among the leading physicists, the theory of relativity was largely accepted by 1911, although not by all. Some opposed it well into the 1920s and 1930s. The Nobel Prize–winning physicists Philip Lenard and Johannes Stark, for example, were not only hostile to theoretical physics in general, but they were also anti-Semites and, later, committed Nazis. Neither Lenard nor Stark, to put it mildly, had any regard for Einstein. In 1911, Max Laue, Max Planck's favorite student, published the first and a very good book on relativity, *Das Relativitätsprinzip*.[29] In that same year, Arnold Sommerfeld, a prominent physicist from Munich, described Einstein's theory of relativity "as one of the secure possessions of physics."[30] In 1910, the faculty of Prague University recommended that Einstein be offered a faculty position for his "epoch-making" work on his theory of relativity. Einstein accepted the position in 1911.[31]

In 1917, Einstein began receiving nominations for the Nobel Prize in physics. However, the road from the nominations to the prize itself was tortured. Einstein's name dominated the list of nominees in 1920, one year after the prediction of the general theory was confirmed by Eddington, but the Nobel Committee for Physics rejected him. No member of the committee approved of the theory of relativity and they further doubted the validity of Eddington's experiment. "Einstein must never receive a Nobel Prize even if the whole world demands it," said a leading member of the physics committee.[32] In the face of many nominations for Einstein, the physics committee, determined as they were to avoid honoring Einstein, decided to award no physics prize in 1921.

As nominees were being considered for the 1922 prize, Einstein had a champion on the Nobel Committee for Physics in the person of Carl Wilhelm Oseen. Oseen knew that committee members were emotionally opposed to relativity, so he developed a strategy. Einstein's particle theory of light, proposed in the March paper, was widely rejected by physicists and could not be the basis of the prize. Oseen argued that Einstein's theory of the photoelectric effect, however, which appeared in the March paper and which had been verified precisely by Robert Millikan, was worthy of the prize. Oseen's strategy included promoting Niels Bohr for the prize at the same time. When the committee voted on September 6, 1922, they agreed to give the 1921 physics prize to Einstein for the photoelectric effect and the 1922 prize to Bohr.

The physics committee made its recommendation to the full Swedish Academy, which had the final say. Some academy members were worried about what the reaction would be for not awarding Einstein the prize for the theory of relativity, but the opposition against relativity was strong. In the end, the academy accepted the physics committee's recommendation and awarded the 1921 Nobel Prize to Einstein "for his services to Theoretical Physics, and especially for his discovery of the law of the photoelectric effect." But there was a stipulation: Einstein was given the prize on the condition that he make no mention of relativity. He was told that his Nobel lecture had to be on the photoelectric effect.

Einstein was unable to attend the formal Nobel Prize ceremony in December 1922. Later, he arranged to give his lecture in the summer of 1923, not in Stockholm, but in Gothenburg. Sitting in the front row of the auditorium in Gothenburg's Lisenberg amusement park was Gustav V, the King of Sweden, who wanted to learn about relativity. For his Nobel lecture, Einstein talked about the theory of relativity after all.

Since 1905, the special theory of relativity has been experimentally tested from right to left, from back to front, and from top to bottom. The theory has met every challenge. Moving clocks *do* run slow, the speed of light *is* the speed limit recognized by Nature, the momentum of an object *does* depend on the inertial reference system it is in, the relativistic Doppler effect *has* been verified, and the list goes on. The tenets of special relativity are tested daily in laboratories where particles are accelerated to speeds approaching that of light. Always, the theory stands firm.

Today the special theory of relativity has taken its place as a super theory. Any physical theory must incorporate Einstein's relativity theory. Quantum mechanics, per se, took its final form in 1927. But it was not really final because it was not compatible with Einstein's theory of relativity. The final step was taken in 1928 when Paul Dirac merged quantum mechanics and relativity. Out of that merger came properties of the electron and aspects of nature, like antimatter, that did not come out of quantum mechanics itself. The power and range of quantum mechanics were enhanced by its compatibility with relativity. In this sense, relativity completed quantum mechanics.

Relativity is a super theory in another sense: its influence beyond physics is substantial. Relativity stimulates and influences philosophical discourse. Epistemology—how we know—is affected by Einstein's relativity. Similarly, idealism, realism, and materialism are discussed differently now than they were before 1905. Artists, writers, and poets have been inspired by the theory of relativity and by its creator. Science routinely affects the larger culture through technology, but physics, which seeks to understand the structural footings of Nature, often influences culture through the novelty of its concepts. The theory of relativity— both special and general—is replete with profound and fascinating ideas and, as a result, has become a part of our culture.

Electrodynamics motivated Einstein's June paper, but it was the kinematic part that has, over the decades, attracted attention. This is partly because electric and magnetic fields are rather abstract ideas and are not a common part of everyday experience. By contrast, the length of objects, clocks, time, simultaneity, time intervals, and speed are routine parts of our lives. We bring a well-established understanding to our notions of length and time. We believe we *know* what simultaneity means. And that is why the public response to the special theory of relativity and its consequences was, and is, one of incredulity among both amateurs and professionals. Even Hendrik Lorentz, one of the greatest physicists of the late nineteenth and early twentieth centuries, certainly understood Einstein's relativity, but he could not quite bring himself to accept the conclusions that resulted from Einstein's June paper. In lectures he gave at the Teyler Foundation in Haarlem in 1913, Lorentz said,

> According to Einstein, it has no meaning to speak of motion relative to the aether. He likewise denies the existence of absolute simultaneity . . . It is certainly remarkable that these relativity concepts, also those concerning time, have found such a rapid acceptance . . . As far as this lecturer is concerned, he finds a certain satisfaction in the older interpretations, according to which the

aether possesses at least some substantiality, space and time can be sharply separated, and simultaneity without further specification can be spoken of . . . Finally, it should be noted that the daring assertion that one can never observe velocities larger than the velocity of light contains a hypothetical restriction of what is accessible to us, [a restriction] which cannot be accepted without some reservation.[33]

All of us, physicists and nonphysicists alike, get comfortable with our common-sense ideas and hate to give them up.

The twentieth century began with three great theories of physics. Today there are five. Two revolutions occurred in physics during the century that fundamentally altered the way physicists view the world. Quantum mechanics changed the way physicists describe the world of atoms and molecules. Einstein's theory of relativity changed the space-time stage on which the drama of Nature unfolds. Looking back to 1905, there is one mistake that Einstein made in May of that year. In his letter to Besso, his close friend, he identified *one* paper, his March paper, as his revolutionary paper. He was wrong. There were *two* revolutionary papers: March, on the particle nature of light and June, on the theory of relativity.

Does the Inertia of a Body Depend on Its Energy Content?

A. Einstein

The paper was received by the editor of the German journal *Annalen der Physik* on September 27, 1905, and published that same year in volume 18, pages 639–641.

Albert Einstein on January 2, 1931, during his third stay in the United States. When Einstein demonstrated in his September paper that energy and mass were equivalent, he may have felt somewhat like he looks in this picture . . . a little impish?

The Most Famous Equation

Galileo, one of the founders of modern science, said that the language of Nature is mathematics. What Galileo said many centuries ago still holds true today. Physics strives to identify the invisible laws that determine the patterns of the visible world. The whole subject is built upon a few surprisingly simple basic laws of Nature. Because they are simple, they can be rendered mathematically. Mathematics requires simplicity of expression; therefore, the basic laws are expressed in terms of mathematics. Even a rudimentary understanding of the invisible laws of physics can inspire wonder: their simplicity masks their range, belies their power, and gives no inkling of the exquisite details they divulge about the diverse actions of Nature.

That Nature's physical laws are simple does not mean that Nature itself is simple. Typically the simplicity of a physical law hinges on the fact that it represents an idealization or an approximation. For example, Galileo asserted that two objects, a light one and a heavy one, when released together from some height, strike the ground at the same time. Well, not exactly, Galileo admitted. Nevertheless, he continued, if the air were removed and the heavy and light objects fell through a vacuum, they would indeed strike the ground together. Air is part of Nature and at some level it must be taken into account, of course, yet idealizing the fall of an object by removing the complication of air is the way Galileo caught a glimpse of a simple law in operation.

The power to apply the basic laws of physics is derived from their mathematical expressions. Bring the dozen or so basic laws together with their manifold applications and the mathematical equations multiply. Physics is replete with equations, simple and complicated. However, some equations enjoy a prominence because their reach is longer and their embrace of Nature is more intimate. And there are a few equations that have acquired symbolic significance.

Each of the five grand theories of physics can boast many wonderful mathematical equations. However, suppose we were asked to select just one equation, one *simple* equation, to symbolize each great physical theory. Which five equations would appear? Opinions may differ, but here are five equations that would be present on many if not most lists:

1. Newtonian Mechanics $F = ma$
   This is Newton's second law, the most basic law of motion. This equation states that when a force F acts on a mass m it accelerates with an acceleration a in the same direction as the applied force.
2. Thermodynamics $S = k \ln \Omega$
   This is what Einstein referred to as Boltzmann's Principle. It relates the entropy S of a state to the logarithm of the probability of the state Ω. Boltzmann's constant is k. A high probability state is a high entropy state.
3. Electromagnetism $c = 1/(\varepsilon_0 \mu_0)^{1/2}$
   Coming out of Maxwell's basic equations of electromagnetism is the speed of light c in terms of an electric constant ε_0 and a magnetic constant μ^0.
4. Relativity $E = mc^2$
   Einstein's famous equation relates energy E and mass m through the square of the speed of light c.
5. Quantum Mechanics $E = h\nu$
   This is sometimes called the Planck equation. It gives the energy E of a particle of light with frequency ν. The letter h represents Planck's constant.

Each of these equations represents a major physical theory. Each of these simple equations has many conceptual layers that, when unpacked, tell a long story. Only one of these equations, however, is known to people across all professions, in all occupations, and in all walks of life throughout the world. That one equation is Einstein's simple equation $E = mc^2$. Not all may understand the particulars of his theory, but they *do* know that Einstein's equation stands for something very important. Whatever the level of understanding people bring to it, Einstein's equation, presented for the first time in his September 1905 paper, has become a part of world culture.

The Setting

What motivated Einstein to write his September 1905 paper? Or, more generally, what motivated him to write any of his 1905 papers? Certainly, Einstein was troubled by the contradictory ideas of light as continuous and matter as discontinuous, yet existing side by side. Although this contradiction did not attract the attention of other physicists in 1905, it did motivate the thinking of Einstein and prompted him to write the revolutionary March paper. Certainly, the formal asymmetries that emerged from moving charges and magnets were unacceptable to Einstein and, to some extent, provoked his thoughts. The special theory of relativity resolved these asymmetries. For Einstein, contradictions and asymmetries were a signal that something was deeply wrong.

Contradictions and asymmetries, as Einstein saw them, were issues of form. What about unresolved experimental puzzles? Did they bother Einstein? When Einstein wrote his March paper, the photoelectric effect was unexplained; when he wrote his May paper, Brownian motion was unexplained; and when he wrote his June paper, the ether had failed to reveal itself even to the most exacting experiments. Certainly the lack of evidence for the ubiquitous ether was tormenting the minds of Einstein's contemporary physicists. Did these physical phenomena, awaiting explanations, motivate Einstein? Einstein had read Poincaré's 1902 book, so he knew about the photoelectric effect and Brownian motion. He actually applied his particle theory of light to the photoelectric effect and resolved all the puzzles associated with that phenomenon that were vexing his contemporaries. But did the photoelectric effect really motivate Einstein or was it just an obvious way to marshal support for his revolutionary particle-of-light theory which, for him, resolved the continuity-discontinuity conundrum?

It is impossible to give a definitive answer to these questions. It is best to take Einstein at his word and understand his motivations in terms of what he gave us—contradictory points of view, asymmetries, logical difficulties, and so forth.

The motivation behind the September paper appears to be more obvious, but do appearances tell the whole story? In a letter to Conrad Habicht, written during the summer of 1905, Einstein writes:

> A consequence of the study on electrodynamics [the June paper] did cross my mind. Namely, the relativity principle, in association with Maxwell's fundamental equations, requires that the mass be a direct measure of the energy contained in a body; light carries mass with it. A noticeable reduction of mass would have to take place in the case of radium. The consideration is amusing and seductive; but for all I know, God Almighty might be laughing at the whole matter and might have been leading me around by the nose.[1]

This letter was most likely the sole portent of Einstein's September paper, which soon followed it. Einstein's words to Habicht suggest two routes by which Einstein might have been brought to the September paper. First, and perhaps the most likely, he came to recognize that the mass-energy relationship was an inevitable consequence of his theory of relativity. After all, that is what he says to Habicht. But with extraordinary prescience Einstein says something more: he mentions radium . . . radioactive radium.

Radioactivity was discovered by Henri Becquerel in 1896 and in the summer of 1905, the underlying physics of radioactivity remained a mystery. The atomic nucleus, the source of radioactivity, was not discovered until 1911. What *was* known in 1905 were some experimental facts about the phenomenon of radioactivity. Specifically, it was known that radioactive atoms suddenly emit highly energetic particles: in some cases rather massive α-particles, in other cases less massive β-particles, and in still other cases, radiation in the form of γ-rays. In all these cases, however, these radioactive ejections possessed energy. Where did the energy come from? Did Einstein reflect on radioactivity and did he wonder how an atom could suddenly erupt and eject an energetic α-particle? Was it the phenomenon of radioactivity that prompted Einstein to examine once again his relativity theory to see if still another unexpected consequence might be forthcoming?

Although no definitive answer is available, Einstein certainly knew about radioactive radium. It could have stimulated a line of thought that resulted in his September paper. Or perhaps he simply enjoyed further pondering his wonderful theory of relativity to see if still more bounty could be buried in it. One way or another, the September paper was the result.

Einstein's September paper is qualitatively different than his other 1905 papers. On the one hand, the equation $E = mc^2$ seemed to come out of the blue. No one anticipated it. There was no recognized need for it. On the other hand, the equation could be derived directly from Einstein's June paper.

Einstein submitted his paper on the special theory of relativity in the last few days of June. Then, at the end of September, he submitted his paper that linked mass and energy. The September paper does not break new ground in the way his June paper did. Quite the contrary. The September paper built upon the June foundations. In the June paper Einstein developed a set of novel ideas, the truth of which was further explored in the September paper. According to the premises of his June paper, E had to equal mc^2. Had Einstein not derived $E = mc^2$ in September, another physicist would likely have done so sometime later. The equation is just a logical consequence. That does not diminish its revolutionary character. Like the contraction of length or the slowing of clocks, the idea that two physical concepts, mass and energy, are the same, that mass is "a direct measure of the energy contained in a body," is one more consequence of the special theory of relativity that challenges our common sense.

The September 1905 Paper

"The results of an electrodynamic investigation published by me recently in this journal," Einstein begins, referring to his June paper, "lead to a very interesting conclusion, which shall be derived here." Einstein then provides a succinct summary of the June paper: "The laws governing the changes of state of physical systems do not depend on which one of the two coordinate systems in uniform parallel translation relative to each other these changes of state are referred to (principle of relativity)."[2]

The September paper is the shortest of Einstein's 1905 papers—only three pages. It could have been the final section of the June paper—it would have made a spectacular conclusion. In both the June and September papers, the reader is called upon to imagine events as seen by observers in two inertial coordinate systems—one at rest and the other in uniform motion. When the results of measurements in the two coordinate systems are compared, bizarre results emerge.

From start to finish, Einstein takes us through three steps. First, he asks, imagine light waves.[3] Light waves (or light particles) carry energy, as was demonstrated in Einstein's analysis of the photoelectric effect. It is the light's energy, transferred to the electrons in the metal's surface, that ejects electrons from the metal's surface. The energies of Einstein's imagined light waves are measured relative to two inertial coordinate systems. First, the energy is measured in coordinate system 1 and next in coordinate system 2, which is moving with a constant velocity v relative to system 1. Using the theory of relativity developed in the June paper, Einstein records the energy of light waves from the perspective of both inertial coordinate systems.

In the second step Einstein considers an object that is at rest with respect to one inertial coordinate system, but is moving with a constant velocity v with respect to a second inertial coordinate system. As always, observers can be identified with each coordinate system.

To appreciate Einstein's third step, we need a definition. In the second step Einstein introduced an object. Let us assume it has a mass of m. The first observer stands next to the object and sees it at rest ($v = 0$). The second observer sees the object in motion with a velocity v, which means that for this observer the object has energy of motion, which is called kinetic energy. The second observer sees a moving object with a kinetic energy of $\frac{1}{2} mv^2$. However, is the object moving with a velocity v to the east, or is the second observer moving with a velocity v to the west? In either case, the second observer sees the object moving with a velocity v and possessing a kinetic energy equal to $\frac{1}{2} mv^2$. It is better to simply say that there is relative motion between the second observer and the object and that either the object is moving relative to the observer or that the observer is moving relative to the object.

Following these two steps, Einstein proceeded to put the two

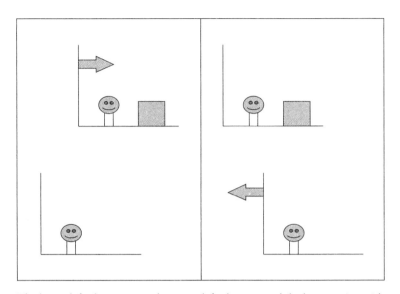

The lower left observer sees the upper left observer and the box moving with a constant velocity v toward the right. The upper right observer sees the lower right observer moving at the same constant velocity v toward the left. Since the two inertial coordinate systems are equivalent, there is no physical way to establish which inertial coordinate system, upper or lower, is actually moving. All one can say is that the lower and upper coordinate systems are moving relative to each other. The lower observer sees the box moving with a velocity v and having a kinetic energy $\frac{1}{2}mv^2$ while the upper observer sees the box as stationary.

steps together in step three. Assume that the object in the previous step emits two light waves in exactly opposite directions. If each light wave has the energy $E/2$, then the total energy carried away by the two light waves is E. Since the light waves originate in the object and since they carry away a total energy of E, the energy of the object must decrease by the amount E. Einstein examines this light-emission process from the perspective of two observers: one at rest with respect to the object (the upper observer in the illustration) and a second for whom the object is in motion (the lower observer in the illustration). According to the principle of relativity, the laws of physics are the same in all inertial coordinate systems, so the conservation of energy can be employed in both the coordinate system at rest and the system in motion. Einstein then shows that for the second observer (lower observer) watching the moving object, the energy loss E due to the emission of light is seen as a decrease in the object's kinetic energy. How can the kinetic energy, $\frac{1}{2}mv^2$, decrease? Either by a decrease in the object's velocity, v, or by a decrease in its mass, m. But since the velocity can be identified with either the object or the observer (as the illustration shows), the decrease in the object's kinetic energy cannot be due to a decrease in velocity. That means the object's mass must decrease. In fact, Einstein shows that the object's kinetic energy decreases by the amount $\frac{1}{2}(E/c^2)v^2$, which means that the object's mass decreased by the amount E/c^2. This result can be written in a more revealing form:

mass lost $= m_{lost} = $ (Energy lost$/c^2$) $= (E_{lost}/c^2)$

or,

$$E_{lost} = m_{lost}c^2$$

or more simply,

$$E = mc^2.$$

Einstein then continues:

From this equation it follows directly: If a body releases the energy E in the form of radiation, its mass decreases by E/c^2. Since obviously here it is inessential that the energy withdrawn from the body happens to turn into energy of radiation rather than into some other kind of energy, we are led to the more general conclusion: The mass of a body is a measure of its energy content; if the energy changes by E, the mass changes in the same sense by E/c^2.[4]

The standard form of Einstein's famous equation, $E = mc^2$, puts the emphasis on energy. If his words above are read carefully, however, Einstein might well have thought of it differently. He writes, "The mass of a body is a measure," which suggests an emphasis on mass; in fact, in terms of the words he used, Einstein expressed his equation as

$$m = E/c^2,$$

which puts mass, m, in the prominent position of the equation. Was this deliberate? Was this another example of Einstein's amazing prescience? More later.

Einstein begins the September paper by referring to the June paper. The "results of an electrodynamics investigation," writes Einstein, "lead to a very interesting conclusion." The conclusion is the famous $E = mc^2$. But is it true? Or, as Einstein wondered, was "God Almighty leading [him] around by the nose"? In typical fashion, he connects his theoretical results with ways to test them. Einstein concludes this paper by suggesting a possible test to determine whether energy and mass were related as he predicted. The radioactive element radium had been discovered by Marie Curie seven years earlier, in 1898. Since an atom of radium gives off energy in its radioactive-decay process and since that energy might well come at the expense of mass, Einstein suggested that it might be possible to test his result "using bodies whose energy content is variable to a high degree (e.g., salts of radium)."

Einstein's final sentence in the September paper considers the practical outcome of his theory: "If the theory agrees with the facts, then radiation transmits inertia [mass] between emitting and absorbing bodies."[5] Since radiation is nothing but energy and since inertia is mass, Einstein's words could have been "energy transmits mass." What does it mean? According to his equation, energy and mass are one. Because they appear to be so different, their oneness is stunning. How does the speed of light, squared, figure in this equation?

Energy is a concept that is hard to define. It comes in many different forms. It comes as electromagnetic energy from the Sun, it comes as electrical energy (or nuclear) from a power company, it comes as chemical energy from gasoline, it comes as thermal energy from a hair dryer, it comes as potential energy from a wound-up spring clock, it comes as kinetic energy from an avalanche, it comes as acoustic energy from someone's mouth, and it comes as geothermal energy from Earth's interior. Energy easily slips from one form to another. Energy is everywhere and energy makes things happen. Energy is the intangible part of Einstein's famous equation.

Mass is linked to materiality. Every material object has mass. Like energy, mass is not easy to define. There are two kinds of mass, each of which can be demonstrated. Kick a big rock and the resulting pain is a demonstration of the rock's inertial mass. Objects with a large inertial mass strongly resist being moved. Hold a bag of sugar at arm's length and the sugar's gravitational mass demonstrates itself. The sugar's mass is attracted by Earth's mass (gravity at work), and it requires an effort to keep the sugar from responding to that attraction. Mass is the tangible part of the September equation.

The equation $E = mc^2$ connects intangibility and tangibility and, by making them equivalent, joins them as one. Energy and mass *are* different and it is the speed of light that brings them together.

There are at least three ways to explain the presence of c^2 in the equation that links E and m. First, the c^2 is there because of the June paper, that is, because of the two principles of the special theory of relativity. Start with those two principles, demand that these two principles remain valid for observers in all inertial coordinate systems, follow the logic wherever it leads, and accept the consequences as they emerge. One consequence is length contraction and another is $E = mc^2$. Viewed this way, the equality $E = mc^2$ is the direct consequence of the larger theory. This is not very satisfying, though true. The question still lingers: Why is the speed of light in the equation?

The second way to explain the presence of c^2 is as a conversion factor. A conversion factor converts one unit into another. For example, miles and kilometers are different units of length or distance. If you want to convert a mile into a kilometer, you must multiply the mile by the conversion factor 1.609. The c^2 is what converts mass units into energy units. Just as a volume expressed in quarts cannot be equated to a distance expressed in miles, so a mass expressed in kilograms cannot be equated to an energy expressed in joules. Kilograms and joules are incompatible units. A conversion factor is needed.[6] Here again, the explanation falls short of satisfying.

A third way to explain the presence of the c^2 in Einstein's equation is, I believe, more satisfying. Begin with the recognition that for centuries men and women have sought to understand Nature. Recognize further that everyday experiences, in their totality, provide the prism through which people observe and interpret Nature. These same common experiences have had decisive influences on the way people describe and understand Nature. Next, acknowledge lessons repeatedly learned from past experiences; namely, explanations based on common experiences cannot automatically be extended beyond those experiences. Finally, accept what Einstein taught us in the September paper (which has since been amply verified); namely, at the basic level, Nature does *not* distinguish between energy and mass. Humans distinguish between energy and mass, but Nature does not. Even more, humans have made mass into something very different than energy. This difference is demonstrated by the observation that humans have created a kilogram as the unit for mass and a joule as the unit for energy—two distinctly incompatible units. If, however, the objective is to describe Nature accurately, humans must accept Nature on its terms and find a way to rationalize the difference between our concept of mass and our concept of energy. The factor c^2 does this. Multiply m by c^2, and, de facto, energy and mass become what Nature deems them to be: one and the same.

Einstein himself described the mass energy equation a little differently:

> It follows from the theory of relativity that mass and energy are both different manifestations of the same thing—a somewhat unfamiliar conception for the average man. Furthermore, $E = mc^2$, in which energy is put equal to mass multiplied with the square of the velocity of light, showed that a very small amount of mass may be converted into a very large amount of energy . . . the mass and energy in fact were equivalent.[7]

Einstein's June paper brought space and time together. His September paper brought energy and mass together. Space and time seem totally different, and yet Einstein showed that to describe Nature accurately, the two must be joined. Mass and energy also seem to be two completely unrelated things. They do not look alike, they do act alike, but at their root, Nature tells us that they are one. That is the conclusion of Einstein's September paper.

The Response

In 1907, Einstein was invited to write a review paper about relativity theory. In this paper he reaffirmed his argument for the equivalence of mass and energy when he stated: "[I]nertial mass and the energy of a physical system appear . . . as things of the same kind . . . It seems far more natural to consider any inertial mass as a reserve of energy."[8] In this same paper Einstein acknowledges that the suggestion he made earlier in his September 1905 paper, the suggestion to test the mass-energy equivalence by means of an experiment with radioactive radium, had been investigated by the physicist J. Precht. For some reason, Einstein does not quote Precht but quotes Max Planck's summary of Precht's experimental results. Specifically, if the energy coming from 226 grams of radioactive radium originates completely from mass, the 226 grams of radium would decrease by 0.000012 gram per year. Since this mass decrease is too small to detect, Einstein's equation cannot be tested by means of radium. But his original intuition, that the place to look for evidence supporting mass-energy equivalence was some kind of transformation inside the atom, was on target. It just took some time.

The mass-energy equivalence did not come into its own until the 1930s when nuclear physics became an active area of physical research. When physicists established that the atomic nucleus consisted of protons and neutrons, a new question had to be answered: How can the protons, vigorously repelling each other because of their positive charges, be held together in the snug confines of the nucleus? The answer is provided by Einstein's equation. When protons and neutrons come together to form a nucleus, each gives up a trifling amount of its mass. In other words, the mass of an assembled atomic nucleus is less than the sum of the masses of its unassembled parts. Where does the mass go? The missing mass becomes the energy that binds the nucleus together. If Δm represents the mass lost by particles when they assemble together in a nucleus, then the energy holding the nucleus together is Δmc^2, or the binding energy E equals Δmc^2. Once the particles are assembled in a nucleus and bound together by an energy E, the same energy is required to take the nucleus apart particle by particle. Every atom in the universe has a nucleus which testifies to the equivalence of mass and energy.

In the same 1907 review paper, Einstein once again anticipates things to come. After he acknowledges that the mass-energy equivalence could not be tested with radium, he writes, "However, it is possible that radioactive processes will be detected in which a significantly higher percentage of the mass of the original atom will be converted into the energy of a variety of radiations than in the case of radium."[9] Einstein was wrong; it was not a radioactive process, but his instincts, focused on the atomic interior, were eerily prescient.

At the end of 1938, nuclear fission was discovered. When a uranium nucleus is struck by a neutron, it splits into two large parts, a few neutrons, and a lot of energy. Upon examination, it is found that the total mass of the pieces after fission is less than the mass of the pieces before fission. The energy produced comes from the loss of mass. The fission process is the basis of the atomic bomb. It was the recognition that the fission process held the potential for a devastating weapon that persuaded Einstein to sign a letter in 1939 to President Roosevelt urging him to initiate a program to determine the feasibility of a nuclear weapon.

Large atomic nuclei can break into two relatively large parts and, in the process, release energy. Small atomic nuclei can join together into larger nuclei and, in the process, release energy. The September equation $E = mc^2$ explains both fission and fusion.

Einstein's little equation also makes the Sun shine. At the core of all stars, hydrogen fuses with hydrogen to form helium and, in the process, mass becomes energy. In time, the energy generated in a star's core works its way to the surface and, for stars like the Sun, maintains a surface temperature of about 5,800 degrees. For many years physicists sought to explain how the Sun produced its energy. It was from Einstein's September equation that we learned what fires the stars.

Einstein's equation came before its time. It was not until the

1930s that the atomic nucleus required the mass-energy equivalence. Later, as twentieth-century physicists probed deeper and
deeper into the structure of matter, the September equation
served its purpose.

In accelerator laboratories around the world, Einstein's equation is a staple. The equivalence of mass and energy must be
taken into account in the design of accelerators and it is essential
when particle collisions are analyzed. In the gigantic detectors
surrounding the location where the high-energy particle beams
collide, the oneness of mass-energy is witnessed over and over
again as masses vanish into energy and masses reappear from
energy.

With the discovery of antimatter, Einstein's equation became
indispensable. The first antiparticle, the positron, was discovered
in 1932 and in later years, the discovery of other antiparticles
followed. When a particle meets its antiparticle, they annihilate
each other and the two masses become energy. Again, it works
both ways. The energy of a high-frequency photon can suddenly
become the source of a particle-antiparticle pair. Creation—energy to mass—and annihilation—mass to energy—go back and
forth indiscriminately. As creation and annihilation processes
dramatically demonstrate, Nature makes no distinction between
mass and energy.

One of the more incredible scientific advances of recent years
has resulted from the connection made between two worlds: the
world of the smallest and the world of the largest. Earthbound
physicists examine the details of the elementary particles in accelerator laboratories, while astronomers and cosmologists are intellectually transported to the far reaches of the cosmos to examine how the universe came into existence. New data, pouring out
of the Hubble Space Telescope on almost a weekly basis, have
pushed back the thoughts of cosmologists closer and closer to the
origin of the universe itself: the big bang. As cosmologists have

sought to understand the cosmic environment and events that occurred in the earliest seconds following the big bang, they have needed information gleaned from Earth-based accelerator laboratories. As the worlds of the largest and smallest have come together, Einstein's equation has been an indispensable tool. The big bang could not be understood without $E = mc^2$. During the first few seconds following that singular event, mass and energy danced a solo as annihilations and creations of particles set the stage for the universe we now inhabit. At the end of those early seconds, the universe had cooled to the point where the dance between mass and energy had become a *pas de deux:* energy and mass *could* change partners because they were linked by Nature, but the ease with which they switched identities back and forth had lessened and they began to assume the appearances that would prompt humans to make them separate entities: human mass and human energy.

Ask any physicist to identify the most important concept of physics and the answer is likely to be "energy." And for good reason. As physics has developed, no concept enjoys the range of applicability that is enjoyed by the energy concept. Perhaps this is why physicists choose to write Einstein's September equation as $E = mc^2$, where the energy, E, is the subject of the equation. Einstein did not write his equation this way; rather, he made mass the subject of the equation: $m = E/c^2$. Did Einstein's acute intuition prompt him to write it that way?

Now in the twenty-first century, physicists have at their disposal a highly developed model, the standard model, that identifies the basic building blocks of matter. The model is refined to the point that questions once ignored now demand answers. One such question is: where do the basic particles get their masses? More specifically, why is the mass of the proton what it is rather than something else? Advances in the further understanding of matter may hinge on answering such a question. We might begin by writing Einstein's September equation the way he did, $m = E/c^2$, and ask: Is the mass of the proton and other basic particles to be found in a particular form of energy? Does Nature have pockets of energy that happily become the proton and other basic building blocks of the material world?[10]

It is certainly a stretch to suggest that in 1905 Einstein anticipated the need to explain the mass of the basic particles. The basic particles were unknown in 1905. But the electron, discovered in 1897, was known. The electron had an unknown mass, but it was thought to be a small mass. Did Einstein wonder, as he finished his September paper, whether perhaps the electron's small mass might be understood in terms of energy, causing him to express his equation in terms of mass and not energy? It would be risky to underestimate Einstein's powerful intuition that enabled him "to scent out that which was able to lead to fundamentals and to turn aside from everything else."

Einstein's September equation, $E = mc^2$, is unique. It plays a starring role in those elegant abstractions that stimulate the minds of those at the frontiers of science while remaining a familiar part of modern culture.

The Foundation of the General Theory of Relativity

A. Einstein

The Quantum Theory of Radiation

A. Einstein

Quantum Theory of Single Atom Ideal Gases

A. Einstein

Can Quantum-Mechanical Description of Physical Reality Be Considered Complete?

Albert Einstein, Boris Podolsky, and Nathan Rosen

Einstein's paper on the general theory of relativity was published in 1916 in *Annalen der Physik*, volume 49. "The Quantum Theory of Radiation" was published in 1917 in *Physikalische Zeitschrift*, volume 18. His paper concerning Bose-Einstein statistics and condensate, "Quantum Theory of Single Atom Ideal Gases," was published in 1924 and 1925 in *Sitzungberichte der Preussischen Akademie der Wissenschaften zu Berlin*. And the Einstein, Podolsky, and Rosen paper, "Can Quantum-Mechanical Description of Physical Reality Be Considered Complete?" appeared in 1935 in the American journal *Physical Review*, volume 47.

Albert Einstein in 1954. He appears fully at peace. He died on April 18, 1955, approximately one year after this picture was taken.

Beyond 1905

This book celebrates Einstein's greatest year, 1905. Einstein's accomplishments in that year were quickly recognized as unusual as well as provocative and they established Einstein as a physicist to be watched. Einstein's 1905 publications, however, neither brought him immediate fame nor showered him with job offers. Despite the fact that he was publishing on a monthly basis in Germany's leading physics research journal, *Annalen der Physik,* no universities competed for Einstein's services in 1905. So he continued as a civil servant at the Bern patent office until October 1909, four years after his last 1905 paper was published, when he acquired a faculty position at the University of Zurich. In the same month, Einstein attended his first physics conference in Salzburg.

After his appointment at the University of Zurich, professional opportunities came rapidly to Einstein. In 1911, Emperor Franz Joseph appointed Einstein full professor at the Karl-Ferdinand University in Prague. He accepted the appointment and on April 3, 1911, he and his family arrived in Prague. Less than one year later, he was appointed professor of theoretical physics at the Federal Institute of Technology (ETH) in Zurich and in July 1912, Einstein and family returned to Zurich. By this time, Einstein was in demand. In July 1913, Max Planck and Walther Nernst, two leading German scientists, traveled to Zurich to visit with him. Their purpose was to discuss the possibility of his accepting a professorship at the University of Berlin. Following their visit to Zurich, Planck, Nernst, and others nominated Einstein for membership in the Prussian Academy of Sciences, a high honor. Later in 1913, Einstein accepted the Berlin offer, which included a professorship with no teaching obligations and the directorship of the Kaiser Wilhelm Institute for Physics. In the spring of 1914, Einstein arrived in Berlin.[1] In 1913, eight years after one of the most incredible years in the history of science, Einstein was received with full honors in the top echelons of German physics.

By 1920, he was an international celebrity—a rare position for a professor of physics. Einstein was fortunate. In 1905 he was a young man of twenty-six, still an outsider to the organized physics profession. His thinking had not been influenced by consensus views on the proper way to do physics, on what is possible or impossible in physics, or what is important or unimportant in physics. In 1905 he had not only a job that he enjoyed, but also one that allowed his mind to range freely over ideas that interested him. In 1905 he was in love, but he did not have family responsibilities that would have competed for his attention.

Perhaps most significant, however, is that in 1905, the subject of physics was brimming with potential. Discoveries had been made during the final years of the nineteenth century that had disrupted the serenity of many physicists and had jarred them out of their relaxed notion that physics was approaching its final form. These discoveries shook the underpinnings of physics in virtually all areas of the subject. In 1895 an unknown form of penetrating radiation was discovered, which were called X-rays. In 1896, it was discovered that a chemical element was itself the source of an unknown radiation stemming from a process that came to be called radioactivity. In 1897, a negatively charged particle was discovered whose properties hinted that the atom may have smaller parts. This particle was the electron. Finally, in late 1900, the quantum was born, which forced physicists to think in new ways. The events from 1895 to 1901 were total surprises and, with the exception of X-rays, standard physical theory provided no explanations. New ideas were needed. In the intellectual environment of 1905, Einstein thrived.

As a young man, Einstein was indeed fortunate, but in the longer term, he was unfortunate. When Einstein was in his mid-forties, still relatively young, quantum mechanics was created in response to an array of questions about matter at the level of atoms. Physicists used quantum mechanics to answer the old questions and identify new ones. With successes leading to more successes, quantum mechanics demonstrated its amazing power. With the formalism of quantum mechanics, physicists could calculate atomic properties and put their calculations to the test through experiments. The abstractness of quantum mechanics and its decisive break with earlier physics challenged and excited physicists.

Quantum mechanics quickly came to dominate the thoughts and the work of front-line physicists in the mid to late 1920s. The theory of quantum mechanics, however, as it was conceptualized after 1927, approached Nature in terms of probabilities that Einstein could not accept. Einstein was a realist. He believed that Nature was real, that atoms were real, and that atomic properties were real. "Proper" physics had the potential for determining the exact physical properties of atoms. Quantum mechanics was at odds with Einstein's realist views; from the perspective of quantum mechanics, his views were in contradiction with Nature. Although Einstein's colleagues did not hesitate to use the tools provided by quantum mechanics, Einstein responded by pursuing his own dream theory, the unification of gravitation and electromagnetism, and that pursuit effectively took him out of the mainstream of physics for the remaining years of his life. One can only wonder what additional contributions Einstein may have made to physics had he been able to accept the physical and philosophical underpinnings of quantum mechanics. For more than half of Einstein's professional life, the intellectual environment that dominated physics was hostile to his basic beliefs. In the intellectual environment of physics after 1927, Einstein stood apart.

1905: Einstein's Base

Einstein was prolific. His scientific papers first appeared in 1901 and continued until (and including) the year he died. Einstein was consumed by physics. By no means, however, was he a one-dimensional man. He wrote books for the general reader, he participated a bit in world affairs, he was a musician of sorts and enjoyed music, and he was a copious correspondent. Nonetheless, physics is what he lived for and what dominated his waking hours.

Although Einstein enjoyed a long career, he never had another year that compared with 1905. His 1905 papers were not only important in and of themselves, but they formed a base that influenced essentially all of Einstein's subsequent work. The base he built in 1905 gave Einstein both substantive and procedural footings.

The 1905 papers were substantive in both their range and their depth. The subjects that Einstein addressed in 1905 were not confined to one narrow topic. On the contrary, four separate conceptual domains provided the framework for the 1905 papers: What is light? What is the nature of matter? Is classical thermodynamics valid? Are physical laws valid for all observers? Collectively these four domains embraced a major part of physics, and hence his papers influenced the entire discipline. Moreover, in none of the papers did Einstein solve a problem that, once solved, could be put on the shelf and forgotten. Rather, he proposed and developed physical ideas whose range extended far beyond the pages of his particular papers. The intellectual richness of the ideas he presented in the 1905 papers was such that Einstein and others came back to these same ideas singly and in combination again and again. In fact, the implications of Einstein's 1905 papers are still with us and they continue to stimulate and perplex physicists.

The March paper on the quantum theory of light stands alone. It broke more decisively with tradition than any of the other papers. Furthermore, Einstein's theories concerning light abuse the fundamental tenets of common sense every bit as much as relativity theory changes our understanding of space and time or of mass and energy. In March, Einstein effectively made light both a wave and a particle. In 1909 Einstein wrote a paper with the title, "On the Development of Our Views Concerning the Nature and Constitution of Radiation," in which he states:

> Once it was recognized that light exhibits the phenomena of interference and diffraction, it seemed hardly doubtful any longer that light is to be conceived as a wave motion . . . However . . . it is undeniable that there is an extensive group of facts concerning radiation that shows that light possesses certain fundamental properties that can be understood far more readily from the standpoint of Newton's emission theory of light [light as particles] than from the standpoint of the wave theory. It is therefore my opinion that the next stage in the development of theoretical physics will bring us a theory of light that can be understood as a kind of fusion of the wave and emission [particle] theories of light.[2]

The dual nature of light became a core consequence of quantum mechanics and the paradox of duality remains with us.

In the same 1909 paper quoted above, Einstein brings into his consideration of light his equation $E = mc^2$. Once again, and a little more clearly, Einstein shows that an object loses mass when it loses energy through the emission of light. He writes:

Thus the inertial mass of a body decreases upon emission of light. The energy emitted must be reckoned as part of the body's mass . . . The theory of relativity has thus changed our views on the nature of light insofar as it does not conceive of light as a sequence of states of a hypothetical medium [the ether], but rather as something having an independent existence just like matter.[3]

Einstein never stopped thinking about the implications of his March 1905 paper.

The April paper, Einstein's dissertation on molecular dimensions, was the precursor to the May paper on Brownian motion; the latter is the more important. For Einstein, the May paper may have been as much about statistical fluctuations as it was about Brownian motion. Einstein's explanation of the random motion of particles suspended in a liquid was based on fluctuations that occur in the motions of the liquid molecules. After 1905, fluctuation theory was a powerful weapon in Einstein's arsenal.

The June paper, which presented the special theory of relativity to the world, was especially important to Einstein. It is undoubtedly one of the greatest papers in the history of physics. The September paper, noteworthy as it was, must be considered as the completion of Einstein's June work. Throughout the years leading up to 1916, the special theory was rarely out of Einstein's conscious thoughts.

In addition to a substantive base, the papers of 1905 also provided Einstein with a procedural base. Einstein started his March and June papers with contradictory images and new physics resulted when Einstein resolved the contradictions. In the March and June contradictions, there is irony. In March Einstein could be seen as rejecting Maxwell's electromagnetic wave theory of light in favor of a particle theory; then in June, three months later, Einstein could be seen as contradicting himself by rescuing

Maxwell's electromagnetism from internal contradictions. Einstein used contradictions because they opened avenues of reflective thought that led him to issues deeper than the apparent contradiction. Anyone can see contradictions, but Einstein saw in the contradictions what others did not. All physicists saw light as continuous waves and matter as discontinuous particles, but no physicist other than Einstein saw in this a contradiction.

Einstein was driven to simplify and unify. These two principles dominated his approach to physics. In his March 1905 paper, Einstein brought radiation and matter together by making radiation, like matter, particle in nature. At the same time, Einstein recognized that the facts, interference and diffraction, fit beautifully with the wave theory of light. So, driven by his need to bring disparate views together, Einstein called for "a kind of fusion" of the wave and particle theories of light.

The June paper exudes simplicity. The entire special theory of relativity is derived from two simple principles, the concepts of space and time, which are unified and brought out of their Newtonian isolation. A world with absolute space existing apart from and independent of absolute time was turned into a world where space and time are joined. Energy and mass, never before regarded by any physicist as having anything to do with each other, were made one as a result of Einstein's September paper.

Beyond 1905

How many great papers did Einstein produce? The answer depends, of course, on how the count is made and who is doing the counting. Most knowledgeable people would place at least three of Einstein's 1905 papers on the list: the March, May, and June papers. After 1905, there is one paper that would be at the top of every list; then there are another half-dozen other papers that would be selected by most. Einstein set a high standard in 1905.

Measured against that, Einstein's great papers after 1905 were few in number. For my list, I have selected one topic and four papers, all of which had a significant impact on physics.

The Equivalence Principle

There is no single paper in which Einstein presented what came to be called the equivalence principle. Einstein referred to it explicitly and implicitly in several papers over the period from 1907 to 1911, and he used it in dramatic ways. The equivalence principle unifies disparate concepts. As such, this principle is one of those vintage Einsteinian insights that had a tremendous influence on his own thinking and on physics. It can be expressed in two ways, both of which are true, but both provoke different sorts of thought.

Statement 1: Uniform gravitation cannot be distinguished from uniform acceleration.
Statement 2: Gravitational mass and inertial mass are one and the same.

One image brings the essence of Statement 1 into focus. The illustration shows Einstein's ever-present observer in an enclosed cabin. In one case the cabin is sitting on Earth and, if the cabin is small, gravity is uniform throughout the cabin; in the other case the cabin is located in gravity-free space, far from other masses, and it is accelerating in the direction shown. The equivalence principle states that the enclosed observer has no way to distinguish between these two cases. The observer, standing on a bathroom scale, weighs 184 pounds in both cases. No difference. The observer drops a ball, and the ball falls to the floor in exactly the same way in both cases. The observer flips a coin and the coin follows a parabolic path in both cases. Just as observers moving uniformly relative to each other cannot perform an experiment

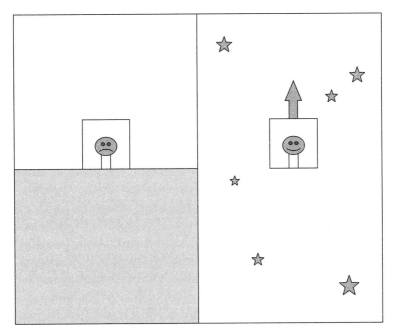

One observer (left) is in a cabin based on Earth. A second observer (right) is in a space-based cabin accelerating as indicated by the arrow. Einstein's equivalence principle states that there is no way these two observers can determine which of them is on Earth and which is in space.

to determine who is moving and who is at rest, so there is no experiment that the observer can perform that will distinguish between the accelerated cabin and the cabin on Earth. Einstein says that these two very different situations are equivalent.

Statement 1 can be illustrated by a different image. Imagine a small cabin falling toward Earth. Next imagine a small cabin far from Earth in a gravity-free space. These two situations, free-fall and zero gravity, are equivalent. There is no way to distinguish between free-fall and zero gravity. To experience zero gravity, we would have to jump out of a tall building or ride in a freely falling elevator. For the brief moments of free-fall (before smashing into the ground) we would experience zero gravity.

The second statement derives from experience: we experience mass in two distinctly different ways. An object's *inertial* mass is a measure of the resistance it gives to being moved, specifically, being accelerated. A baseball is easier to throw than a bowling ball because the inertial mass of the baseball is smaller than that of the bowling ball. An object's gravitational mass is a measure of the attraction between it and another mass. Earth and a bowling ball mutually attract each other because of their gravitational masses. In his equivalence principle, Einstein asserts that these two masses, the inertial mass and the gravitational mass, are exactly the same.

For decades the best physicists observed gravitation and acceleration every day. For decades, many of the same physicists absolutely recognized a similarity between gravitation and acceleration, but it was Einstein who saw in the connection an equivalence that represented something profound: the equivalence of gravitation and acceleration would, with Einstein's guidance, become warped spacetime.

Where did the equivalence principle come from? In 1907, Einstein had what he called "the happiest thought of my life," which I quoted in a previous chapter:

> I was sitting in a chair in the patent office at Bern when all of a sudden a thought occurred to me. "If a person falls freely he will not feel his own weight." I was startled. This simple thought made a deep impression on me. It impelled me toward a theory of gravitation.[4]

Einstein expressed this idea, which formally became the equivalence principle, in several different places at different times including 1907 and 1911.

The General Theory of Relativity—1916

The special theory of relativity was two years old when, in 1907, Einstein was asked by Johannes Stark to review the scientific literature and to report to readers what had been written on the subject of special relativity since June 1905. This paper, titled "On the Relativity Principle and the Conclusions Drawn from It," was an important undertaking. It required Einstein to step back and look at the theory afresh and to write about it in a more pedagogical style. This review article brought Einstein face to face with the limitations of the special theory and it represents an early step, if not the first step, toward the general theory. The last major section of this 1907 paper carries the title, "Principle of Relativity and Gravitation." Clearly, in this paper, Einstein had advanced beyond his theory of June 1905.

Even though Einstein recognized the limitations of the special theory, it took him several years of concentrated effort to think through the implications of a simple but commanding idea—the equivalence principle—that he believed would lead to a generalization of his June 1905 paper. In 1912, four years before he reached his objective, Einstein told his fellow physicist in Munich, Arnold Sommerfeld, that compared to the general theory, the special theory was "child's play."[5] The creation of the general theory was an intellectual tour de force and many regard the outcome as the all-time greatest product of pure thought.[6]

To appreciate Einstein's challenge, let us think again about coordinate systems. The "absolute space" of Newton came to be filled with the ether that was the coordinate system of choice, a preferential coordinate system, Nature's coordinate system. But Einstein, for one, thought that a coordinate system is simply a matter of convenience, merely a means to describe Nature, and is not a fundamental part of Nature itself. Inertial coordinate systems, systems that move uniformly (at a constant speed and in a straight line) relative to each other, had the place of honor in both Newton's physics and in the special theory of relativity. One of the axioms of the special theory is that all inertial coordinate systems are equivalent as far as the laws of physics are concerned. If an observer in any one inertial coordinate system puts herself in the shoes of an observer in another inertial coordinate system, their descriptions of Nature would fit perfectly with the same laws of physics. But why should *inertial* coordinate systems be singled out as special? Singling out uniform-motion coordinate systems from nonuniform-motion, that is, accelerated coordinate systems represented an asymmetry that Einstein could not accept. The validity of the laws of physics should not be limited to a particular type of coordinate system. That was Einstein's objective: to remove the special theory's restriction to inertial coordinate systems and to develop a theory that he believed (and he believed it without qualification) to be true; namely, that *all* coordinate systems are equivalent and the basic laws of physics formulated in any one coordinate system apply without alteration in all other coordinate systems. Bringing *all* coordinate systems together was a simplification and a unification that appealed to Einstein.

Einstein's starting point—his simple and commanding idea— was the equivalence principle. As Einstein said in a 1921 talk in London, "The general theory of relativity owes its existence . . . to the empirical fact of the numerical equality of the inertial and gravitational masses of bodies."[7] All physicists recognized a connection between inertial and gravitational masses, but Einstein recognized much more: he recognized that to generalize the special theory, the equivalence of inertial and gravitational masses dictated that gravity would have to be a core part of the new theory.

As I stated earlier, the equivalence principle equates uniform gravity and a uniformly accelerated coordinate system. From this equivalence principle alone, consequences of general relativity can be qualitatively anticipated. For example, the general theory quantitatively predicts that a ray of light should be bent by the gravitational influence of a massive object. If there is an equivalence between gravity and acceleration, then light should bend in the accelerated coordinate system of the cabin. This is shown in the illustration on p. 140. In the same illustration, a light source is present on the floor of the accelerating cabin, shining toward the ceiling. During the same instant that the light travels from the floor to the ceiling, the ceiling has moved away from the light source with the result that the crests arriving at the ceiling are farther apart than they were upon leaving the floor and the light appears to shift toward the red end of the visible spectrum.[8] If this is true in an accelerated coordinate system, it must also be true in a gravitational environment. And it is. In 1960, Robert V. Pound and Glen Rebka showed that light leaving Earth is redshifted. These strange effects and others became quantitative with the completion of the general theory.

The concepts of space, time, and gravitation are dramatically changed by the general theory. Gravity affects time: clocks run slower in the strong gravity near the Sun than do identical clocks in the weaker gravity far from the Sun. The sources of gravity, masses, also affect space: space is warped by masses. (And since mass and energy are equivalent [the September paper], energy warps space.) A large mass warps space in such a way that another mass is drawn into the space surrounding the large mass. In this sense, the general theory of relativity has replaced the gravitation force, as described by Newtonian physics, with warped spacetime.

For many years after 1916, the general theory was recognized for its pristine intellectual beauty, but it had few experimental applications. This is no longer true. The study of astronomical objects like black holes and neutron stars requires general relativ-

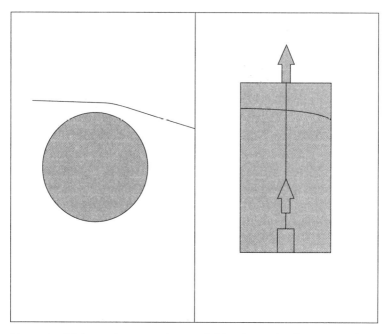

The path of light passing a massive star (left) is diverted or bent. In an elevator accelerating upward (right), light leaves the left wall of the elevator. As it travels across the elevator, the elevator itself moves upward with the result that the light strikes the right wall at a lower point than it departed from. In the accelerating elevator the light appears to bend just as it bends passing a star. On the floor of the elevator is a light source directing light upward. Since the ceiling is running away from the light, the light seen by the ceiling is red-shifted.

ity. The general theory is also a necessary tool for understanding the origin of the universe—the big bang. Gravitational waves, predicted by the general theory, are believed to emanate from various astronomical objects. A neutron star, for example, orbiting another star, would be a source of gravitational waves. The search is under way to detect these waves.

The special theory of relativity quickly became a necessary tool in experimental laboratories of physics. It took longer for the more exotic general theory to enter the laboratory, but it has done so and it has also become a necessary tool in many areas of physical research.

The Quantum Theory of Radiation—1917

Einstein wrote several papers on quantum issues. The first, of course, was his March 1905 paper in which he resurrected Newton's long-defunct particle model for light and presented it in quantum terms. The light quantum was a constant challenge to Einstein. In May 1911, Einstein wrote to his friend Michele Besso, "I no longer ask whether these quanta really exist. Nor do I try to construct them any longer, for I know that my brain cannot get through in this way."[9] The light quantum haunted Einstein's thoughts for most of his life. In 1916 and 1917, Einstein continued his thinking on the quantum nature of radiation and three papers resulted. These three papers were a continuation of Einstein's "heuristic point of view" presented in March 1905. Einstein's quantum theory of light, rejected by virtually all physicists, was considerably strengthened by the 1916–1917 papers, which testified to Einstein's strong belief that light had a particle nature. The last in this series of three, appearing in 1917, is one of Einstein's most famous papers.[10]

In typical Einstein fashion, he begins the 1917 paper by pointing out a "formal similarity" that is "too striking."[11] In this paper, Einstein considered atoms in a bath of radiation emitting and absorbing particles of light. By this means he was able to give a new derivation of Max Planck's famous blackbody radiation law (without the arbitrary assumptions that Planck had to make) and, in the process, showed as a necessary condition that when atoms go from one energy state to another, a light quantum is either emitted or absorbed. In this way, Einstein connected Planck's law with Bohr's model of the hydrogen atom. Einstein clearly recognized that he had strengthened his light-particle position:

> This derivation [the Planck radiation law] deserves consideration not only because of its simplicity, but especially because it appears to clarify the processes of emission and absorption of radiation in matter, which is still in darkness for us . . . [A]s a result of this our simple hypothesis about the emission and absorption of radiation acquire [sic] new supports.[12]

In 1917, the origin of an atom's spectrum was conceptualized in terms of Bohr's atomic model. An atom's absorption spectrum, represented by dark spectral lines, resulted when atoms absorbed energy from radiation and moved from a lower to a higher energy state. An atom's emission spectrum, represented by bright spectral lines, occurred when atoms spontaneously emitted radiation and moved from a higher to a lower energy state. Einstein added to this picture by making a provocative supposition. He assumed not only that atoms could *spontaneously* emit light, but also that they could be induced or *stimulated* by light to emit light; specifically, light of frequency v could stimulate an atom to emit light of frequency v and, in the process, move from a high energy state to a lower energy state.

For decades Einstein's idea of *stimulated* emission sat dormant. Then in 1954 came the MASER, the precursor to LASER (the acronym for *L*ight *A*mplification by the *S*timulated *E*mission of *R*adiation). The laser is based on Einstein's assumption of stimulated emission: one light particle enters a system of atoms, stimulates emission, and two light particles emerge; hence, one light particle is doubled to two light particles—the amplification. Charles H. Townes, one of the inventors of the laser, wrote of Einstein's 1917 paper, "Albert Einstein was the first to recognize clearly, from basic thermodynamics, that if photons can be absorbed by atoms and lift them to higher energy states, then it is necessary that light can also force an atom to give up its energy and drop to a lower level. One photon hits the atom, and two come out . . . The result is called stimulated emission and results in coherent amplification."[13]

At the end of his 1917 paper, Einstein writes that his analysis "makes the quantum theory of radiation almost unavoidable . . . The weakness of the theory lies . . . in not bringing us closer to a union with the wave theory."[14] Yet in spite of Einstein's powerful arguments, arguments that effectively strengthened Bohr's atomic model, Bohr and other physicists did not accept Einstein's particle theory of light for another five years.

Bose-Einstein Statistics and Condensate—1924, 1925

On June 4, 1924, an unknown Indian physicist from the University of Dacca, Satyendranath Bose, sent Einstein a short manuscript with the title, "Planck's Law and the Light Quantum Hypothesis." Bose had sent this manuscript earlier to the prestigious journal *Philosophical Magazine* and it was rejected. He asked Einstein for help. Einstein was so favorably impressed with the manuscript that he translated it into German and sent it to the journal *Zeitschrift für Physik* with an appended note saying he thought the paper by Bose represented an important advance.

In his manuscript, Bose had successfully derived Planck's blackbody radiation law by a statistical approach based on the assumption that light quanta were indistinguishable from each other. Einstein recognized immediately that the same approach could be applied to indistinguishable atoms. Bose's paper and two papers by Einstein were the origin of Bose-Einstein statistics.[15]

The elementary particles that form the building blocks of matter fall into one of two classes: bosons and fermions. The photon, gluon, and pion are examples of bosons; the electron, proton, and quarks are examples of fermions. Bose-Einstein statistics applies to bosons, whereas Fermi-Dirac statistics, developed by Enrico Fermi and Paul Dirac, applies to fermions. Bosons and fermions are different in that an unlimited number of bosons can come together and occupy the same quantum state. By contrast, only one fermion can occupy a particular quantum state.

In his 1925 paper, Einstein made a prediction that bosons could exhibit an unusual behavior. Because many bosons can crowd together in the same quantum state, it is possible, under the right conditions of temperature, density, and so on, for bosons to coalesce and condense into a new state of matter. In describing a boson gas, Einstein used similar words: "I maintain that . . . [a] separation is effected; one part condenses, and the rest remains a 'saturated gas.'"[16] This state of matter, called the Bose-Einstein condensate, was observed for the first time on June 5, 1995, by Eric Cornell and Carl Weiman at the National Institute of Standards and Technology in Boulder, Colorado. The study of BE condensates is now one of the more active areas of physical research. Seventy years after Einstein predicted a new form of matter, it was discovered.

The Einstein, Podolsky, and Rosen Paper—1935

The last of Einstein's papers that I will discuss sets the stage for an active and extremely provocative area of future research.[17] The purely imaginary experiment described by Einstein, Boris Podolsky, and Nathan Rosen has been actualized and now real experiments inspired by this paper are exposing fascinating and unbelievable properties of the real world.

In 1933, Einstein attended a lecture in Brussels given by Léon Rosenfeld. The lecture was on quantum mechanics and Einstein paid close attention. Following the lecture, Einstein, in the audience, directed the ensuing discussion to the meaning of quantum mechanics. He asked Rosenfeld:

> What would you say to the following situation? Suppose two particles are set in motion towards each other with the same, very large, momentum, and that they interact with each other for a very short time when they pass at known positions. Consider now an observer who gets hold of one of the particles, far away from the region of interaction, and measures its momentum; then, from conditions of the experiment, he will obviously be able to deduce the momentum of the other particle. If, however, he chooses to measure the position of the first particle, he will be able to tell where the other particle is. This is a perfectly correct and straightforward deduction from the principles of quantum mechanics; but is it not paradoxical? How can the final state of the second particle be influenced by a measurement performed on the first, after all physical interaction has ceased between them?[18]

Two years later, in 1935, the question Einstein posed to Rosenfeld formed the basis of the Einstein-Podolsky-Rosen paper, known throughout physics as the EPR paper.

The issue raised by Einstein's question and the EPR paper can be stated succinctly and dramatically. When an atom goes from a higher to a lower energy state, it can emit two photons, which fly away from the atom in opposite directions. If at some time in the future a measurement is made on one photon, thereby changing it, the second photon, halfway across the universe, will change instantaneously. That something happening in one place can instantly affect what happens in another place Einstein called "spooky." It violated his view of causality. It also raised basic issues about the adequacy of the quantum mechanical description of Nature.

The EPR paper hit the devoted quantum mechanicians "as a bolt from the blue."[19] Bohr, always responsive to any challenge by Einstein to quantum theory, spent six weeks developing a response to the Einstein paper. In Bohr's response, he concluded that the EPR paper did not necessitate any changes to quantum mechanics and, soon thereafter, physicists largely forgot the EPR paper. For over three decades it resided in the dusty archives of physics: a curious paper by a great physicist who, unlike all other physicists, did not accept the tenets of quantum mechanics.

Then John S. Bell wrote a paper in 1966 that brought the EPR paper out of the archives and into the mainstream. Just as Cornell and Wieman's discovery of the Bose-Einstein condensate started a research industry, so John Bell's paper stimulated research activities by both theoretical and experimental physicists throughout the world.

The two photons emitted by the atom just described are said to be entangled. "Entanglement," a term coined by Erwin Schrödinger in 1935, is one of the more important discoveries of the last century and it has its roots in the EPR paper. As the term "entanglement" implies, two parts of a system separated by a vast distance may appear to be independent, but they are not. The two parts are entangled. In 1972, John Clauser and Stuart Freedman showed that entanglement is an actual phenomenon. Experiments by Alain Aspect in 1983 in France began the experimental study of entangled states in earnest.

Einstein, Podolsky, and Rosen's 1935 paper and John Bell's 1966 paper have presented physicists with uncomfortable alternatives. Einstein was a realist and believed in what is now called locality—what happens in a particular locality cannot immediately influence what happens in another locality. Einstein in fact insisted on locality: "But on one supposition we should, in my opinion, absolutely hold fast: the real factual situation of the system S_2 is independent of what is done with the system S_1, which is spatially separated from the former."[20] What has emerged from subsequent science, however, is that locality and quantum mechanics cannot both be right. If we accept quantum mechanics, we must give up locality, or vice versa. Had he lived another twenty years, Einstein would have been deeply troubled by the alternatives that physics laid before him.

Einstein's Legacy

Would the world now be different if Albert Einstein had never lived? Could we ask the same question with regard to Claude Monet or Wolfgang Amadeus Mozart? What is the relative impact of a legendary figure of science compared to a legendary figure of art or music?

Although both art and science are human activities, they are thought about in different ways. Monet's *Palazzo da Mula* and Mozart's *Die Zauberflöte* are regarded as wondrous acts of creativity. Had Monet not lived, the world would be different because the *Palazzo da Mula* never would have been painted; had Mozart not lived, the world would be different because the opera *Die Zauberflöte* never would have been composed. By contrast, had Einstein not lived, the world would be no different. His special theory of relativity, a response to the intellectual environment of 1905, inevitably would have been created by someone else. Framed this way, art becomes a highly creative activity with the fingerprints of an artist personalizing every painting and composition, and science becomes an intellectual activity driven by events—shared by the larger science community but external to the scientist.

Framed this way, however, the natures of both art and science are obscured. Art is also driven by events external to the artist. Impressionist painters, who lived and worked during a culturally revolutionary period, the late nineteenth and early twentieth centuries, influenced each other. Although each brushstroke expressed the individuality of painters such as Monet, Auguste Renoir, and Edgar Degas, collectively the brushstrokes left a canvas that articulated a new theory of art. When Monet painted *Palazzo da Mula,* he was driven by external influences that had a determining influence on the outcome, but in the end, it was a painting like no other, a Monet masterpiece, a monument to human creativity. In 1883, Renoir said, "I had wrung Impressionism dry."[21] Soon thereafter the ideas that drove Impressionism and that had inspired a great school of art were superseded.

Mozart lived and worked during the late eighteenth century. His contemporaries included Franz Joseph Haydn and Ludwig van Beethoven. Composers of this era were strongly influenced by the acoustical nature of the concert halls available for performances as well as the musical range and mechanical efficiency of the musical instruments available to performers. Although each individual note of Mozart's *Die Zauberflöte* was his and his alone, the score of this famous opera bears the signature of the classical period. Like Monet, Mozart was driven by external events, but in the end, the final score of the opera *Die Zauberflöte* was like no other, a product of high creativity. By the early to mid-nineteenth century, musicians threw off the constraints imposed by the classical period and composed more personal and emotional music.

As in the visual arts and in music, science always has a context and a community. Einstein was influenced by his contemporaries as well as by the state of physics in 1905 and beyond. It is indeed likely that if Einstein had not created the special theory of relativity, someone else would have created something equivalent to Einstein's theory. However, just as paintings by Claude Monet and Edouard Manet belong to the same genre and yet are each unique, we can imagine a theory by Einstein and a similar theory by, say, Poincaré, motivated by the same concerns. The theories would have similarities, but each would be unique. Einstein's theory would be distinguished from Poincaré's theory by the starting point adopted, the conceptual path followed, the assumptions made, and the form of its final outcome. Each theory would be a unique product of human creativity.

A few artists have such a distinctive style that their art, be it painting or music, stands apart. The same can be said for a few scientists. Perhaps no scientist had a more distinctive style than Albert Einstein. The general theory of relativity, as Einstein created it, is such a masterpiece, physics of the rarest kind. In time, another physicist would have been motivated, either for experimental or theoretical reasons, to extend Einstein's special theory of relativity to noninertial coordinate systems and thereby generalize it to all coordinate systems; in time, another physicist may have recognized something deeper in the connection between inertial and gravitational masses; in time, gravitational forces may have been seen in terms of spatial properties. No one but Einstein ever would have put all these elements together in the same simple, harmonious, and elegant way. Just as a composition from the mind of Mozart reveals his artistic uniqueness, so the general theory of relativity reveals Einstein's scientific uniqueness. The general theory, considered by many to be the greatest monument to abstract thought, prompts the same kind of wonder and the same kind of emotion as does an artistic masterpiece. "The equations of general relativity," wrote Stephen Hawking, "are his best epitaph and memorial. They should last as long as the universe." [22]

This book began by recognizing Isaac Newton and Albert Einstein as the two greatest physicists of all time. These two men invite attention. Jacob Bronowski, in his book *Ascent of Man,* writes:

It is almost impertinent to talk of the ascent of man in the presence of two men, Newton and Einstein, who stride like gods. Of the two, Newton is the Old Testament god; it is Einstein who is the New Testament figure. He was full of humanity, pity, a sense of enormous sympathy. His vision of nature herself was that of a human being in the presence of something god-like, and that is what he always said about nature. He was fond of talking about God: "God does not play dice," "God is not malicious." Finally Niels Bohr one day said to him, "Stop telling God what to do." But that is not quite fair. Einstein was a man who could ask immensely simple questions. And what his life showed, and his works, is that when the answers are simple too, then you hear God Thinking.[23]

In 1905, Einstein had a direct line to God's thoughts.

Toward the end of his life, Einstein looked back to 1905 and told a friend, Leo Szilard, "They were the happiest years of my life. Nobody expected me to lay golden eggs."[24] The expectation of "golden eggs" was a consequence of his wondrous year of 1905. The quality and the quantity of groundbreaking work produced by Einstein from March through September have no equal. The year 1905 set a standard for Einstein himself. Only in 1916, after years of intense effort, did he surpass the bar he established for himself in 1905.

The year 1905 set another standard, a standard with wide-ranging influence. Of all human activities, thinking is the single activity that most clearly sets us apart from other life forms. Thinking is what makes us human. Einstein's 1905 is an illustration of the thinking species at its best, the thinking person's standard of greatness.

Notes

Further Reading

Acknowledgments

Index

Prologue The Standard of Greatness

1. Letter from Einstein to Conrad Habicht, May 18 or 25, 1905, *The Collected Papers of Albert Einstein, Volume 5, The Swiss Years: Correspondence, 1902–1914*, trans. Anna Beck (Princeton University Press, 1995), p. 20.

2. Einstein's Kyoto Address, December 1922. Quoted in J. Ishiwara, *Einstein Koēn-Roku* (Tokyo-Tosho, 1977). Also quoted in Abraham Pais, *"Subtle Is the Lord": The Science and the Life of Albert Einstein* (Oxford University Press, 1982), p. 179.

3. Letter from Einstein to Arnold Sommerfeld, October 29, 1912, *Collected Papers*, vol. 5, p. 324.

4. Quoted in Silvio Bergia, "Einstein and the Birth of Special Relativity," in *Einstein: A Centenary Volume*, ed. A. P. French (Harvard University Press, 1979), p. 67.

5. Edwin F. Taylor and John Archibald Wheeler, *Spacetime Physics: Introduction to Special Relativity*, 2nd ed. (W. H. Freeman & Co., 1992), p. iii.

6. Arthur I. Miller, *Einstein, Picasso: Space, Time, and the Beauty that Causes Havoc* (Basic Books, 2001), p. 4.

7. Albert Einstein, "Autobiographical Notes," in *Albert Einstein: Philosopher-Scientist*, ed. Paul Arthur Schilpp (Harper & Brothers, 1959), pp. 15, 17.

8. Quoted in Albrecht Fölsing, *Albert Einstein*, trans. Ewald Osers (Viking, 1997), p. 439.

9. Albert Einstein to Johannes Stark, September 25, 1907, *Collected Papers*, vol. 5, p. 42.

10. Albert Einstein to Michele Besso, November 17, 1909, 1912, *Collected Papers*, vol. 5, p. 140.

11. Fölsing, *Albert Einstein*, p. 548.

12. Philipp Frank, *Einstein: His Life and Times* (Alfred A. Knopf, 1970), p. 110.
13. Ibid., p. 50.
14. Pais, *"Subtle Is the Lord,"* p. 242.
15. Ibid.
16. Hubert Goenner, "Albert Einstein and Friedrich Dessauer: Political Views and Political Practice," *Physics in Perspective* 5: 21–66.
17. Jacob Bronowski, *The Ascent of Man* (Little, Brown, 1973), p. 256.

March   The Revolutionary Quantum Paper

1. Albert Einstein in *The Collected Papers of Albert Einstein, Volume 2, The Swiss Years: 1900–1909,* trans. Anna Beck (Princeton University Press, 1989), p. 87.
2. Albrecht Fölsing, *Albert Einstein: A Biography,* trans. Ewald Osers (Viking, 1997), p. 143.
3. Letter from Einstein to Conrad Habicht, May 18 or 25, 1905, *Collected Papers,* vol. 5, p. 20.
4. Max Laue letter to Einstein, June 2, 1906, *Collected Papers,* vol. 5, pp. 25–26.
5. Charles Coulson Gillispie, *The Edge of Objectivity: An Essay in the History of Scientific Ideas* (Princeton University Press, 1960), p. 456.
6. James Clerk Maxwell, quoted in Gillispie, *Edge of Objectivity,* p. 476.
7. Albert Einstein, "Autobiographical Notes," in *Albert Einstein: Philosopher-Scientist,* ed. Paul Arthur Schilpp (Harper & Brothers, 1959), p. 45.
8. Einstein, *Collected Papers,* vol. 2, p. 86.
9. In his extensive work on the influence of "themes" on the development of science, Gerald Holton has examined the strong influence continuity had on Einstein's thinking. See Holton's *Thematic Origins of Scientific Thought: Kepler to Einstein* (Harvard University Press, 1973), p. 357.
10. Einstein, "Autobiographical Notes," p. 37.
11. Einstein, *Collected Papers,* vol. 2, p. 94.
12. Earlier I represented energy with the symbol ε, but here with the symbol E. What's the difference? The energy of an *individual* quantum is represented by ε; the energy of a *system* with many, many quanta is represented by E.
13. Einstein, *Collected Papers,* vol. 2, p. 97.

14. Ibid., pp. 99–100.
15. Fölsing, *Albert Einstein*, p. 147.
16. Ibid.
17. Robert A. Millikan, "A Direct Photoelectric Determination of Planck's "*h*," *Physical Review* 7: 355–388, p. 355 (1916).
18. Ibid.
19. Robert A. Millikan, *The Electron: Its Isolation and Measurements and the Determination of Some of Its Properties* (University of Chicago Press, 1917), p. 224.
20. Ibid., p. 230.
21. A September 6, 1916, letter from Albert Einstein to Michele Besso, quoted in Abraham Pais, "Einstein On Particles, Fields, and the Quantum Theory," in *Some Strangeness in the Proportion: A Centennial Symposium to Celebrate the Achievements of Albert Einstein,* ed. Harry Woolf (Addison-Wesley, 1980), p. 209.
22. Quoted in Abraham Pais, *Niels Bohr's Times, In Physics, Philosophy, and Polity* (Oxford University Press, 1991), p. 233.
23. Ibid., p. 231.
24. Roger H. Stuewer, *The Compton Effect: Turning Point in Physics* (New York: Science History Publications, 1975).
25. Quoted in Martin, Kline, "Einstein and the Development of Quantum Physics," pp. 133–151 in *Einstein: A Centenary Volume,* ed. A. P. French (Harvard University Press, 1979), p. 133.

April Molecular Dimensions

1. Albert Einstein letter to Michele Besso, January 22, 1903, in *The Collected Papers of Albert Einstein, Volume 5, The Swiss Years: Correspondence, 1902–1914,* trans. Anna Beck (Princeton University Press, 1995), p. 7. When Einstein wrote this letter, he was employed at the Bern patent office. In that environment he may have concluded that a doctorate would be of little benefit.
2. Quoted in Albrecht Fölsing, *Albert Einstein: A Biography,* trans. Ewald Osers (Viking, 1997), p. 75.
3. Maja Einstein quoted in Fölsing, *Albert Einstein*, p. 123.
4. Fölsing, *Albert Einstein*, p. 124.
5. Albert Einstein, "Autobiographical Notes," in *Albert Einstein: Philosopher-Scientist,* ed. Paul Arthur Schilpp (Harper & Brothers, 1959), p. 49.

6. Einstein letter to Michele Besso, March 17, 1903, in *Collected Papers,* vol. 5, p. 11.
7. Einstein, *Collected Papers,* vol. 2, p. 105.
8. Ibid.
9. Abraham Pais, *"Subtle Is the Lord": The Science and the Life of Albert Einstein* (Oxford University Press, 1982), p. 90.
10. Tony Cawkell and Eugene Garfield, "Assessing Einstein's Impact on Today's Science by Citation Analysis," pp. 31–40 in *Einstein: The First Hundred Years,* ed. Maurice Goldsmith, Alan Mackay, and James Woudhuysen (Pergamon Press, 1980), p. 32.

May "Seeing" Atoms

1. Albert Einstein, *The Collected Papers of Albert Einstein, Volume 2, The Swiss Years: 1900–1909,* trans. Anna Beck (Princeton University Press, 1989), p. 123.
2. Einstein, *Collected Papers,* vol. 2, p. 123.
3. Mary Jo Nye, *Molecular Reality: A Perspective on the Work of Jean Perrin* (Elsevier, 1972), p. 11.
4. C. P. Snow, "Einstein," pp. 3–18 in *Einstein: The First Hundred Years,* ed. Maurice Goldsmith, Alan Mackay, and James Woudhuysen (Pergamon Press, 1980), p. 9.
5. Einstein, *Collected Papers,* vol. 2, p. 127.
6. Ibid., p. 134.
7. Ibid., p. 318.
8. Jean Perrin, "Brownian Movement and Molecular Reality," trans. F. Soddy, in Mary Jo Nye, *The Question of the Atom: From the Karlsruhe Congress to the First Solvay Conference, 1860–1911* (Tomash Publishers, 1984), 2nd ptg. 1986, p. 567.
9. Ibid.
10. Ibid., p. 568.
11. Jean Perrin, *Oeuvres scientifiques de Jean Perrin* (Paris: Centre National de la Recherche Scientifique, 1950), p. 218. See also C. P. Snow, "Einstein," pp. 3–18 in *Einstein: The First Hundred Years,* ed. Maurice Goldsmith, Alan Mackay, and James Woudhuysen (Pergamon Press, 1980), p. 9.
12. Perrin, *Oeuvres scientifiques,* p. 567.
13. Albrecht Fölsing, *Albert Einstein: A Biography,* trans. Ewald Osers (Viking, 1997), p. 132.

14. Bernard Pullam, *The Atom in the History of Human Thought* (Oxford, 1998), p. 256.
15. Quoted in Stephen G. Brush, *The Kind of Motion We Call Heat: A History of the Kinetic Theory of Gases in the 19th Century,* vol. 2 (North-Holland Pub., 1976), p. 699.
16. Martin J. Klein, "Fluctuations and Statistical Physics in Einstein's Early Work," in *Albert Einstein: Historical and Cultural Perspectives,* ed. Gerald Holton and Yehuda Elkana (Princeton University Press, 1982), p. 53.
17. Abraham Pais, *"Subtle Is the Lord": The Science and the Life of Albert Einstein* (Oxford University Press, 1982), p. 100.
18. Max Born, "Einstein's Statistical Theories," in *Albert Einstein: Philosopher-Scientist,* ed. Paul Arthur Schilpp (Harper & Brothers, 1959), p. 166.
19. Einstein, *Collected Papers,* vol. 5, p. 130.

June   The Merger of Space and Time

1. Albert Einstein, "Autobiographical Notes," in *Albert Einstein: Philosopher-Scientist,* vol. 1, ed. Paul Arthur Schilpp (Harper & Brothers, 1959), p. 21.
2. Ibid., p. 23.
3. Werner Heisenberg, *Encounters with Einstein* (Princeton University Press, 1983), p. 29.
4. P. A. M. Dirac, "The Excellence of Einstein's Theory of Gravitation," in pp. 41–46 of *Einstein: The First Hundred Years,* ed. Maurice Goldsmith, Alan Mackay, and James Woudhuysen (Pergamon Press, 1980), p. 44.
5. Gerald Holton, "Einstein, Michelson, and the 'Crucial' Experiment," *Isis* 60: 133–197 (1969). Quotation on p. 137.
6. Abraham Pais, *"Subtle Is the Lord": The Science and the Life of Albert Einstein* (Oxford University Press, 1982), pp. 126–127.
7. Ibid., p. 127.
8. Ibid., p. 139.
9. John Stachel, *Einstein's Miraculous Year: Five Papers that Changed the Face of Physics* (Princeton University Press, 1998), p. 112.
10. Arthur I. Miller, *Einstein, Picasso: Space, Time, and the Beauty that Causes Havoc* (Basic Books, 2001), p. 193.
11. Albert Einstein, *The Collected Papers of Albert Einstein, Volume 2,*

The Swiss Years: Writings, 1900–1909, trans. Anna Beck (Princeton University Press, 1989), p. 140.

12. Ibid., p. 141.
13. Ibid., p. 143.
14. Ibid., p. 145.
15. Ibid., p. 149.
16. Ibid., vol. 5, p. 20.
17. Ibid., vol. 2, p. 159.
18. Pais, *"Subtle Is the Lord,"* pp. 149–150.
19. Einstein, *Collected Papers,* vol. 2, p. 171.
20. Quoted in Gerald Holton, *Thematic Origins of Scientific Thought: Kepler to Einstein* (Harvard University Press, 1973), p. 235.
21. Einstein, *Collected Papers,* vol. 2, p. 283.
22. As quoted by Holton in *Thematic Origins.* Also in Einstein, *Collected Papers,* vol. 2, p. 284.
23. Letter from Alfred Bucherer to Einstein, September 7, 1908, *Collected Papers,* vol. 5, p. 83.
24. Christa Jungnickel and Russell McCormmach, *Intellectual Mastery of Nature, Theoretical Physics from Ohm to Einstein, Volume 2, The Now Mighty Theoretical Physics 1870–1925* (University of Chicago Press, 1986), p. 323.
25. See Stanley Goldberg, *Understanding Relativity* (Birkhäuser, 1984).
26. L. Pearce Williams, ed., *Relativity Theory: Its Origins & Impact on Modern Thought* (John Wiley & Sons, 1968), p. 120.
27. Pais, *"Subtle Is the Lord,"* p. 316.
28. Williams, *Relativity Theory,* p. 129.
29. Max Laue, *Das Relativitätsprinzip* (Friedrich Vieweg & Sons), 1911.
30. Jungnickel and McCormmach, *Intellectual Mastery of Nature,* p. 309.
31. Ibid., p. 297.
32. Robert Marc Friedman, *The Politics of Excellence: Behind the Nobel Prize in Science* (Henry Holt & Company, 2001), p. 133.
33. Pais, *"Subtle Is the Lord,"* pp. 166–167.

September   The Most Famous Equation

1. Albert Einstein, *The Collected Papers of Albert Einstein, Volume 5, The Swiss Years: Correspondence 1902–1914,* trans. Anna Beck (Princeton University Press, 1995), p. 21.

2. Einstein, *Collected Papers,* vol. 2, p. 172.
3. Just six months earlier, in his March paper, Einstein argued for a particle model of light. Here he refers to light waves. In 1909, long before light acquired a dual nature—both particle and wave—through the implications of quantum mechanics, Einstein treated light in dualistic terms.
4. Einstein, *Collected Papers,* vol. 2, p. 174. In this quotation, Einstein used the letter L to symbolize energy rather than E, and the letter V rather than c to represent the speed of light.
5. Ibid.
6. The units for mass and energy are incompatible: a joule cannot be combined with a kilogram. However, when mass, in kilograms, is multiplied by speed squared, that is by (meter/second)^2, the product kg m^2/s^2 equals energy units, joules. In this sense c^2 is a conversion factor.
7. *The Quotable Einstein,* ed. Alice Calaprice (Princeton University Press, 1996), p. 183.
8. Ibid., pp. 286–287.
9. Ibid., p. 288.
10. It was a wonderful talk by Frank Wilczek that got me thinking about making mass the subject of Einstein's equation.

Epilogue   Beyond 1905

1. Einstein's family followed a little later, but soon thereafter, Einstein and his wife, Mileva, separated and she returned to Zurich.
2. Albert Einstein, *The Collected Papers of Albert Einstein, Volume 2, The Swiss Years: Writings, 1900–1909,* trans. Anna Beck (Princeton University Press, 1989), p. 379.
3. Ibid., p. 386.
4. Einstein's Kyoto Address, December 1922. Quoted in J. Ishiwara, *Einstein Koēn-Roku* (Tokyo-Tosho, 1977). Also quoted in Abraham Pais, *"Subtle Is the Lord": The Science and the Life of Albert Einstein* (Oxford University Press, 1982), p. 179.
5. Letter from Einstein to Arnold Sommerfeld, October 29, 1912, in *Collected Papers,* vol. 5 (Princeton University Press, 1995), p. 324.
6. The general theory was first published as A. Einstein, "The Founda-

tion of the General Theory of Relativity," *Annalen der Physik,* 49 (1916).

7. Albert Einstein, *Essays in Science,* Philosophical Library, 1934, p. 48. Quoted in Julian Schwinger, *Einstein's Legacy: The Unity of Space and Time* (Scientific American Library, 1986), p. 238.

8. This red shift, called the gravitational red shift, is different from the red shift that is seen in light from distant galaxies. The light from galaxies is red-shifted because the galaxies are moving away from observers on Earth. That motion away from the observer effectively pulls crests a little farther apart and thereby shifts the light toward the red end of the visible spectrum.

9. Letter from Einstein to Michele Besso, May 13, 1911, in *Collected Papers,* vol. 5 (Princeton University Press, 1995), p. 187.

10. Published as A. Einstein, "The Quantum Theory of Radiation," *Physikalische Zeitschrift* 18 (1917).

11. See Henry A. Boorse and Lloyd Motz, eds., *The World of the Atom,* vol. 2 (Basic Books, 1966), p. 888. Einstein's entire 1917 paper appears here in translation.

12. Ibid., p. 889.

13. Charles H. Townes, *How the Laser Happened: Adventures of a Scientist* (Oxford University Press, 1999), p. 13.

14. See Boorse and Motz, eds., *The World of the Atom,* vol. 2, p. 901.

15. A. Einstein, "Quantum Theory of Single Atom Ideal Gases," *Sitzungsberichte der Preussischen Akademie der Wissenschaften zu Berlin,* 1924 and 1925.

16. Quoted in Pais, *"Subtle Is the Lord,"* p. 430.

17. Albert Einstein, Boris Podolsky, and Nathan Rosen, "Can Quantum-Mechanical Description of Physical Reality Be Considered Complete?" *Physical Review* 47 (1935).

18. Léon Rosenfeld, "Niels Bohr in the Thirties: Consolidation and the Extension of the Conception of Complimentarity," in S. Rozental, ed., *Niels Bohr: His Life and Work as Seen by His Friends and Colleagues* (John Wiley & Sons, 1967), pp. 114–136, quotation on pp. 127–128.

19. Ibid., p. 137.

20. Albert Einstein, "Autobiographical Notes," in *Albert Einstein: Philosopher-Scientist,* ed. Paul Arthur Schilpp (Harper & Brothers, 1959), p. 85.

21. Helen Gardner, *Art through the Ages*, 7th ed., revised by Horst De la Croix and Richard G. Tansey (Harcourt Brace Jovanovich, 1980), p. 782.

22. Stephen Hawking, "A Brief History of Relativity," *Time,* vol. 154, no. 27, December 31, 1999, p. 81.

23. Jacob Bronowski, *The Ascent of Man* (Little, Brown, 1973), p. 256.

24. Ibid., p. 254.

Calaprice, A., ed. *The Expanded Quotable Einstein.* Princeton University Press, 2000.
   Einstein had a way with words and these many quotations are highly memorable.

French, A. P., ed. *Einstein: A Centenary Volume.* Harvard University Press, 1979.
   An excellent collection of papers celebrating the 100th anniversary of Einstein's birth.

Greene, Brian. *The Elegant Universe: Superstrings, Hidden Dimensions, and the Quest for the Ultimate Theory.* W.W. Norton, 1999.
   Greene is an excellent writer and is also an active participant in shaping the frontier of the physics he writes about. Einstein's 1905 papers are integral to the elegant universe that Greene discusses.

Hoffmann, Benesh, in collaboration with Helen Dukas. *Albert Einstein: Creator and Rebel.* Viking, 1972.
   Hoffmann worked with Einstein for three years at the Institute for Advanced Study in Princeton, New Jersey. Helen Dukas was Einstein's longtime assistant and secretary. This is an enjoyable read.

Holton, Gerald. *Einstein, History, and Other Passions: The Rebellion against Science at the End of the Twentieth Century.* Harvard University Press, 2000.
   This book is highly recommended. Holton has written at great length about Einstein. The first part of this book, "Learning from Einstein," is a testament to Holton's insights.

Kragh, Helge. *Quantum Generations: A History of Physics in the Twentieth Century.* Princeton University Press, 1999.
   Kragh, a prominent historian of physics, guides readers through

some of the most significant developments of twentieth-century physics.

Miller, Arthur I. *Einstein, Picasso: Space, Time, and the Beauty that Causes Havoc*. Basic Books, 2001.
Miller is an Einstein scholar and this is a worthy book.

Pais, Abraham. *"Subtle Is the Lord": The Science and Life of Albert Einstein*. Oxford, 1982.
As the title suggests, Pais brings together Einstein's science, including many equations, and his life. It is a scientific biography that is highly regarded.

Stachel, John, ed. *Einstein's Miraculous Year: Five Papers that Changed the Face of Physics*. Princeton University Press, 1998.
For those who would like to read the actual 1905 papers of Einstein (in translation), this is the book.

Thorne, Kip. *Black Holes & Time Warps: Einstein's Outrageous Legacy*. W.W. Norton & Company, 1994.
A wonderful book.

Acknowledgments

This book might never have happened. Although I had contemplated writing a book that focused exclusively on Einstein's incredibly productive year, it was my interactions with Michael Fisher at Harvard University Press that convinced me it was worth doing. Hence this book exists. I thank Michael Fisher for his interest and encouragement. I also thank Kate Brick and others at Harvard University Press for their support of and attention to this book.

One of the reviewers to whom Michael Fisher sent the original manuscript did a wonderful job of carefully reading it. I thank that reviewer for seeing some little things and some not-so-little things that needed my attention.

I also thank Albert Einstein. It has been my privilege to see many of the great physicists of the twentieth century in action. But alas, I neither met nor saw Einstein. In writing this book, however, I learned a great deal about his 1905 work and I came to feel uncommonly close to this great physicist. So, if he is somewhere still thinking about his light particles, I thank him for his amazing 1905 papers and for the beautiful way he went about doing his physics.

Finally, I acknowledge my one critic who reads and rereads everything I write. She is not a scientist. She comes out of the heart of the humanities—literature. She represents the audience for whom I write and her critique is indispensable. If she is confused by a paragraph, something is wrong with the paragraph. She is my wife, Diana, and a thank you to her is only a feeble expression of my deep gratitude.

Index

Absolute rest, 80, 87
Aether. *See* ether
Albedaran, 22
American Journal of Physics, 98
American Physical Society, 98–99
Animalcules, 59
Annalen der Physik, 4, 17, 39, 41, 52, 53, 55, 73, 96, 97, 105, 125, 127
Annihilation and creation, 122
Antiatomists, 46–47
Antimatter, 101, 122
Antiparticle, 122
Ascent of Man, 149
Aspect, Alain, 146
Atomism, 46, 68. *See also* Antiatomists
Atoms, 46, 47, 58, 62, 63; average behavior of, 60–61; statistical fluctuation of, 60–61
Avogadro's number, 51, 52, 67

Bancelin, Jacques, 52, 53
Becquerel, Henri, 111
Beethoven, Ludwig van, 148
Bell, John S., 146
Berthelot, Marcellin, 46
Besso, Michele, 11, 37, 39, 43, 47, 84, 95, 96, 103, 141
Big bang, 122, 123, 140

Black holes, 139
Bohr, Niels, 3, 11, 37, 100, 143, 145, 146, 150; rejects light quanta, 37
Bohr's model, 142, 143
Boltzmann, Ludwig, 30, 61, 62, 71
Boltzmann Principle, 31, 71, 108
Boltzmann's constant, 67, 108
Born, Max, 71
Bose, Satyendranath, 143
Bose-Einstein condensate, 125, 143–144, 146
Bose-Einstein statistics, 125, 143–144
Bosons, 143, 144
Bronowski, Jacob, 14, 149
Brown, Robert, 59
Brownian motion, 3, 4, 46, 52, 53, 54, 57–72, 75, 109, 110, 132; and atomism, 58, 62, 69, 71; Einstein's directive to experimentalism, 68–69; Einstein's theoretical result, 67; physical nature of, 59
Bucherer, Alfred, 97
Buridan's ass, 9

California Institute of Technology, 18